The foundations of mathematics

IAN STEWART AND DAVID TALL

The foundations
of mathematics

OXFORD UNIVERSITY PRESS

Oxford University Press, Walton Street, Oxford OX2 6DP
Oxford New York
Athens Auckland Bangkok Bombay
Calcutta Cape Town Dar es Salaam Delhi
Florence Hong Kong Istanbul Karachi
Kuala Lumpur Madras Madrid Melbourne
Mexico City Nairobi Paris Singapore
Taipei Tokyo Toronto
and associated companies in
Berlin Ibadan

Oxford is a trade mark of Oxford University Press

Published in the United States by
Oxford University Press Inc., New York

© Oxford University Press, 1977

First published 1977
Reprinted (with corrections) 1979
Reprinted 1983, 1985, 1987, 1989, 1994

A catalogue record for this book is available from the British Library

Library of Congress Cataloging in Publication Data available

ISBN 0 19 853165 6

Printed in Great Britain on acid-free paper by
Biddles Ltd, Guildford and King's Lynn

TO

PROFESSOR RICHARD SKEMP

whose theories on the learning
of mathematics have been a
constant source of inspiration.

Preface

THIS book is intended for readers in transition from school mathematics to the fully-fledged type of thinking used by professional mathematicians. It should prove useful to first-year students in universities, polytechnics, and colleges; and to sixth-formers who contemplate further study in mathematics. It should also prove of interest to a wider class of reader, those who have a grounding in elementary mathematics (GCE O-level being ample) and seek an insight into the basic material and thought-processes of mathematics.

The word 'foundations', as used in this book, has a broader meaning than it does in the building trade. Not only do we base our mathematics on these foundations: they make themselves felt at all levels, as a kind of cement which holds the structure together, and out of which it is fabricated. The foundations of mathematics, in this sense, are often presented to students as an extended exercise in mathematical formalism: formal mathematical logic, formal set theory, axiomatic descriptions of number systems, and technical constructions of them; all carried out in an exotic and elaborate symbolism. Sometimes the ideas are presented 'informally' on the grounds that complete formalism is too difficult for the delicate flowering student. This is usually true, but for an entirely different reason.

A purely formal approach, even with a smattering of informality, is psychologically inappropriate for the beginner, because it fails to take account of the realities of the learning process. By concentrating on the technicalities, at the expense of the manner in which the ideas are conceived, it presents only one side of the coin. The practising mathematician does not think purely in a dry and stereotyped symbolism: on the contrary, his thoughts tend to concentrate on those parts of a problem which his experience tells him are the main sources of difficulty. While he is grappling with them, logical rigour takes a secondary place: it is only after a problem has, to all intents and purposes, been solved intuitively that the underlying ideas are filled out into a formal proof. Naturally there are exceptions to this rule: parts of a problem may be fully formalized before others are understood, even intuitively; and some mathematicians seem to *think* symbolically. Nonetheless, the basic force of the statement remains valid.

The aim of this book is to acquaint the student with the way that a practising mathematician tackles his subject. This involves including the standard 'foundations' material; but our aim is to develop the formal approach as a natural outgrowth of the underlying pattern of ideas. A sixth-form student has a broad grasp of many mathematical principles, and our aim is to make use of this, honing his mathematical intuition into a razor-sharp tool which will cut to the heart of a problem. Our point of view is diametrically opposed to that where (all too often) the student is told 'Forget all you've learned up till now, it's wrong, we'll begin again from scratch, only this time we'll get it right'. Not only is such a statement damaging to a student's confidence: it is also untrue. Further, it is grossly misleading: a student who really did forget all he had learned so far would find himself in a very sorry position.

The psychology of the learning process imposes considerable restraints on the possible approaches to a mathematical concept. Often it is simply not appropriate to *start* with a precise definition, because the content of the definition cannot be appreciated without further explanation, and the provision of suitable examples.

The book is divided into four parts to make clear the mental attitude required at each stage. Part I is at an informal level, to set the scene. The first chapter develops the underlying philosophy of the book by examining the learning process itself. It is not a straight, smooth path; it is of necessity a rough and stony one, with side-turnings and blind alleys. The student who realizes this is better prepared to face the difficulties. The second chapter analyses the intuitive concept of a real number as a point on the number line, linking this to the idea of an infinite decimal, and explaining the importance of the completeness property of the real numbers.

Part II develops enough set theory and logic for the task in hand, looking in particular at relations (especially equivalence relations and order relations) and functions. After some basic symbolic logic we discuss what 'proof' consists of, giving a formal definition. Following this we analyse an actual proof to show how the customary mathematical style relegates routine steps to a contextual background—and quite rightly so, inasmuch as the overall flow of the proof becomes far clearer. Both the advantages and the dangers of this practice are explored.

Part III is about the formal structure of number systems and related concepts. We begin by discussing induction proofs, leading to the

Peano axioms for natural numbers, and show how set-theoretic techniques allow us to construct from them the integers, rational numbers, and real numbers. In the next chapter we show how to reverse this process, by axiomatizing the real numbers as a complete ordered field. We prove that the structures obtained in this way are essentially unique, and link the formal structures to their intuitive counterparts of part I. Then we go on to consider complex numbers, quaternions, and general algebraic and mathematical structures, at which point the whole vista of mathematics lies at our feet. A discussion of infinite cardinals, motivated by the idea of counting, leads towards more advanced work. It also hints that we have not yet completed the task of formalizing our ideas.

Part IV briefly considers this final step: the formalization of set theory. We give one possible set of axioms, and discuss the axiom of choice, the continuum hypothesis, and Gödel's theorems.

Throughout we are more interested in the ideas behind the formal façade than in the internal details of the formal language used. A treatment suitable for a professional mathematician is often not suitable for a student. (A series of tests carried out by one of us with the aid of first-year undergraduates makes this assertion very clear indeed!) So this is not a rigidly logical development from the elements of logic and set theory, building up a rigorous foundation for mathematics (though by the end the student will be in a position to appreciate how this may be achieved). Mathematicians do not think in the orthodox way that a formal text seems to imply. The mathematical mind is inventive and intricate; it jumps to conclusions: it does not always proceed in a sequence of logical steps. Only when everything is understood does the pristine logical structure emerge. To show a student the finished edifice, without the scaffolding required for its construction, is to deprive him of the very facilities which are essential if he is to construct mathematical ideas of his own.

Warwick   I. S. and D. T.
October, 1976

Contents

PART I

The intuitive background

THE final goal of this book will be to develop a rigorous, logical description of the basic ideas of mathematics, in particular the various number systems and the concept of proof. The first two chapters provide a starting point by discussing in an informal and intuitive way, what the student is likely to know already about these concepts. A clear idea of what it is that we are trying to formalize is very helpful in comprehending the formal process themselves.

In metaphorical terms: we intend to construct a building. In this part of the book we survey the ground, and draw some preliminary sketches of what we hope the final edifice will resemble. Or we intend to grow a plant; so we make sure we have a fertile seed and the soil in which to grow it.

We therefore permit ourselves, in these two chapters, to make use of our existing mathematical ideas, such as the operations of arithmetic, without, worrying about their logical status. Our aim instead is to explore some of the consequences, and to refine our intuition ready for later work.

1

Mathematical thinking

MATHEMATICS is not an activity performed by a computer in a vacuum. It is a human activity performed in the light of centuries of human experience, using the human brain, with all the strengths and deficiencies that this implies. You may consider this to be a source of inspiration and wonder, or a defect to be corrected as rapidly as possible, as you wish; the fact remains that we must come to terms with it.

It is not that the human mind cannot think logically. It is a question of different kinds of understanding. One kind of understanding is the logical, step-by-step way of understanding a formal mathematical proof. Each individual step can be checked but this may give no idea how they fit together, of the broad sweep of the proof, of the reasons which lead to it being thought of in the first place.†

Another kind of understanding is from the overall point of view, from which we can comprehend the entire argument, as it were, in a single glance. This involves fitting the ideas concerned into the overall pattern of mathematics, linking them up with similar ideas from other areas.

The reasons for wanting this overall understanding are not just aesthetic or educational. The human mind has a distressing and seemingly unavoidable tendency to make errors. Errors of fact, errors of judgement, errors of interpretation. In the step-by-step method we might miss the fact that one line is not a logical consequence of preceding ones. Within the overall framework, however, if an error leads to a conclusion which does not fit into the total picture, the conflict will lead us to search for the mistake.

---

† For an amusing and instructive example of a proof given by a sequence of steps, each logical in itself, with the final step being the desired conclusion and yet giving no clue as to why the theorem is true or how it was thought of in the first place, see J. E. Littlewood's *A Mathematician's Miscellany*, page 30.

For instance, given a column of a hundred ten-digit numbers to add up (all positive integers, say), instead of getting the correct answer which we suppose to be 137568304452, we might easily make an arithmetical error and obtain 137568804452. In copying this answer we might make a second error and write 1337568804452. Both of these errors could escape detection. The first would almost certainly need a step-by-step checking of the calculation. The second, however, might be detected because it does not fit into the overall pattern of arithmetic. A sum of 100 ten-digit numbers will be at most a *twelve*-digit number (since $9999999999 \times 100 = 999999999900$) and the final proposed answer has *thirteen*.

It is a combination of step-by-step and overall understanding which has the best chance of detecting mistakes; not just in numerical work, but in all areas of human understanding. The student must develop both kinds in order to fully appreciate his subject and to be an effective practitioner of it. Step-by-step understanding is a fairly easy thing to achieve; you just take one thing at a time and do lots of 'drill' exercises until the idea sinks in. Overall understanding is much harder; it involves taking a lot of individual pieces of information and making a coherent pattern out of them. What is worse is that having developed a particular pattern which suits the material at one stage, new information may arise which seems to conflict. Often this will mean that the new information is erroneous but sometimes it indicates that the existing pattern is defective. The more radical the new information is, the more likely it is that it does not fit, and that the existing overall viewpoint has to be modified.

The student must be prepared for this to happen. When it does, the learning process is invariably accompanied by feelings of confusion and loss of confidence. The aim of understanding new mathematics is to avoid losing confidence for long enough to sort out the confusion. To be able to do this it is helpful to be able to recognize the situation when it occurs. That is what this first chapter is about.

Concept formation

All mathematicians were very young when they were born. This trivial platitude carries a non-trivial corollary: that even the most sophisticated mathematician must have passed through the complex process of building up mathematical concepts. When first faced with a problem or a new concept the mathematician turns it over in his mind, digging into his experiences to see if he has seen something like it before. This exploratory, creative phase of mathematics is anything but logical. It is only when the pieces begin to fit together and the mathematician gets a 'feel' for the concept, or the problem, that a semblance of order comes. Finally there is the polishing phase where the essential facts are marshalled into a neat and economical proof.

As a scientific analogy, consider the concept 'colour'. A dictionary definition of this concept looks something like: *sensation produced in the eye by rays of decomposed light.* One does not try to teach the concept of colour to a child by presenting him with this definition. ('Now, Johnny, tell me what sensation is produced in your eye by the decomposed light radiating from this lollipop . . . ?') First you teach him the concept 'blue'. To do this you show him a blue ball, a blue door, a blue chair, . . . accompanying each with the word 'blue'. You repeat this with 'red', 'yellow', 'green', . . . and after a while he begins to get the idea: you point to an object he has not seen before and he says 'blue'. It is a relatively easy matter to refine this to 'dark blue', 'light blue', and so forth. After a lot of this, to establish the individual colours, you start again. 'The colour of that door is blue. The colour of this pig is red. What colour is that buttercup?' If he says 'yellow' then he is grasping the concept 'colour'. As the child develops and learns scientific concepts he may eventually be shown a spectrum obtained by passing light through a prism. He may learn about the wavelength of light and, as a fully-fledged scientist, be able to say with precision what wavelength corresponds to light of a particular blue colour. His understanding of the concept 'colour' is now highly refined; yet it will not help him explain to a child what 'blue' is. The existence of a precise and unambiguous definition of 'blue' in teims of wavelength is of no use at the concept-forming stage.

It is the same with mathematical concepts. The reader of this book has a large number of mathematical concepts established in his mind: how to solve a quadratic equation, how to draw a graph, how to add up a geometric progression.He has great facility in arithmetical calculations. The aim is to build on this wealth of mathematical understanding and to refine these concepts to a more sophisticated level. To do this we must use examples, drawn from the reader's experience, and on the basis of these examples introduce new concepts. Once these are established they become part of a richer experience upon which we can again draw to work still higher. Although it is certainly possible to build up the whole of mathematics by axiomatic methods starting from the empty set, using no outside information whatsoever, it is also totally unintelligible to anyone who does not already understand the mathematics being built up. The expert can look at a logical construction written in a book and say 'I guess that thing there is meant to be "zero", so that thing is "one", that's "two", . . . this load of junk must be the integers, . . . what's that? Oh, I think I see: it must be "addition" . . .'. The non-expert is faced with an indecipherable mass of symbols. It is never sufficient to define a new concept without giving enough examples to explain what it looks like and what can be done with it. (Of course, the expert is often in a position to supply his own examples, and may not need much help.)

Schemas

A mathematical concept, then, is an organized pattern of ideas which are somehow inter-related, drawing on the experience of concepts already established. Psychologists call such an organized pattern of ideas a 'schema'. For instance, a young child may learn to count ('one, two, three-four-five: once I caught a fish a-live . . .') progressing to ideas like 'two sweets', 'three dogs', . . . and eventually discovers that two sweets, two sheep, two cows have something in common, and that something is 'two'. He builds a schema for the concept 'two' and this schema involves his experience that he has two hands, two feet, last week we saw two sheep in a field, the fish a-live rhyme goes 'one, two, . . .' and so on. It is really quite amazing how much information the brain has lumped together to form the concept, or the schema. The child progresses to simple arithmetic ('If you have five apples and you give two away, how many will you have left?') and eventually builds up a schema to handle the problem 'What is five minus two?'.

If you now ask 'What is five minus six?' the response will be 'You can't do it', or perhaps just an embarrassed giggle that an adult should ask such a silly question. This is because the question does not fit into the child's schema for subtraction: he goes back to 'five apples and take six away' and realizes that it can't be done. At a much later stage, he will have learned the concepts involving negative numbers, and will give the answer 'minus one'. What has happened? His original schema for 'subtraction' has been modified to accommodate new ideas—perhaps by thermometer scales, or the arithmetic of banking, or whatever—and his understanding of the concept changes. During the process of change, he will run into confusing problems (what does minus one apple look like?) which he eventually resolves to his satisfaction (apples don't behave the same way as thermometer readings).

A large part of the learning process involves making an existing schema accommodate new ideas. This process as we have said, is accompanied by a state of confusion. If it were possible to learn mathematics without encountering this problem, life would be wonderful. Unfortunately, the human mind does not seem to work that way ('There is no royal road to geometry')† and the next best thing is to recognize not just the confusion, but its causes. At various stages in the reading of this book the reader will be confused. Sometimes, no doubt, the cause will be the authors' sloppiness; but often it will be the accommodation process at work. This should be welcomed as a sign that progress is being made—unless it persists for too long!

An example

This process can be illustrated by looking at the historical development of mathematical concepts, which is itself a learning process, although it involves many minds instead of one. When negative numbers were first introduced they met considerable opposition ('you can't have less than nothing') yet nowadays, in this financial world of debits and credits, negative numbers are a part of everyday life. More interesting is the development of complex numbers. Leibniz knew that the square of a positive number or of a negative number must always be positive. If i is the square root of minus one, then $i^2 = -1$, so i cannot be a positive or a negative number. Leibniz believed that it should therefore be endowed with great mystical significance: a non-zero number neither less than zero nor greater than zero. This led to enormous confusion, and distrust concerning complex numbers—it persists to this day in some quarters. Complex numbers do not fit readily into many people's schema for 'number', and students often reject the concept when it is first presented. Modern mathematicians look at the situation with the aid of an enlarged schema in which the facts make sense.

Imagine the real numbers as being marked on a line in the usual way:

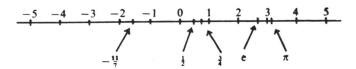

The negative numbers are those to the left of zero, the positive ones to the right. Where does i go? It can't go to the left, it can't go to the right.

† Supposedly said by Euclid to Ptolemy.

The people whose schema does not allow complex numbers must argue thus: this means that it can't go anywhere. There is no place on the line where we can mark i, so it's not a number. Since it's not a number, we certainly can't do arithmetic or algebra using it.

However, we can visualize complex numbers as the points of a plane. The real numbers lie along the 'x-axis', the number i lies one unit above the origin along the 'y-axis', and the number $x + iy$ lies x units along the real line and then y units above it (change directions for negative x or y). The objection to i ('it can't lie anywhere on the line') is

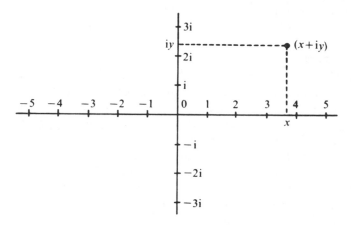

countered by the observation that it doesn't. It lies one unit above the line. The enlarged schema can accommodate the disturbing facts without any trouble. Therefore they cease to be disturbing and the way is clear for a proper understanding of how to use complex numbers in a mathematical problem.

Summary

The above observations on human psychology† have, to a great extent, determined the structure of this book. It will not always be possible to give a precise, dictionary definition of every concept encountered. We may talk about a set being 'a well defined collection of objects', but we will be begging the question, since 'collection' and 'set' mean the same thing.

In studying the foundations of mathematics we must be prepared to become acquainted with new ideas by degrees, rather than starting

† See Skemp [13].

with a watertight definition which can be assimilated at once. Continuing along our path, our understanding of an idea can become more subtle. We can sometimes reach a stage where the original vague definition can be reformulated in a rigorous context ('yellow is the colour of light with a wavelength of 5500 Å'). The new definition, seemingly so much better than the vague ideas which led to its formulation, has a seductive charm. Wouldn't it be so much better to start from this nice, logical definition? The short answer is 'no'. We cannot appreciate the new definition without using our understanding of the concepts involved, and these rest on the original vague ideas, reinforced by our experience. A purely formal approach to mathematics may be extremely edifying to a mathematical logician. It is totally unsuitable for a student. Instead, the student's already considerable knowledge and understanding of mathematics must be reformulated in a way suitable for advanced mathematical thinking. *Then* (and this is the stage we hope to reach by the end of this book) the formal definitions take on real meaning.

Exercises

The following examples are intended to stimulate the reader into considering his (or her) own thought processes and his present mathematical viewpoint. Many of them do not have a "correct" answer, however it will be most illuminating for the reader to write out solutions and keep them in a safe place to see how his opinions may change as he reads the text. The last exercise in the book will be to reconsider the responses to these questions.

1. Think how you think about mathematics. If you meet a new problem which fits into a pattern that you recognize, your solution may follow a time-honoured logical course, but if not, then your initial attack may be anything but logical. Try these three problems and do your best to keep track of the steps you take as you move towards a solution.

(a) John's father is three times as old as John; in ten years he will only be twice John's age. How old is John now?

(b) A flat disc and a sphere of the same diameter are viewed from the same distance, with the plane of the disc at right angles to the line of vision. Which looks larger?

(c) Two hundred soldiers stand in a rectangular array, in ten rows of twenty columns. The tallest man in each row is selected and of these ten, S is the shortest. Likewise the shortest in each column is singled out and T is the tallest of these twenty. Are S and T one and the same? If not, what can be deduced about the relative size of S and T?

Make a note of the way that you attempted these problems, as well as your final solution, if you find one.

2. Consider the two following problems:
 (a) Nine square metres of cloth are to be divided equally between five dressmakers; how much cloth does each one get?
 (b) Nine children are available for adoption and are to be divided equally between five couples; how many children are given to each couple?
 Both of these problems translate mathematically into:

'Find x such that $5x = 9$'.

Do they have the same solution? How can the mathematical formulation be qualified to distinguish between the two cases?

3. Suppose that you are trying to explain negative numbers to someone who has not met the concept and you are faced with the comment:
 'negative numbers can't exist because you can't have less than nothing.'
How would you reply?

4. What does it mean to say that a decimal expansion 'recurs'? What fraction is represented by the decimal $0{\cdot}333\ldots$? What about $0{\cdot}999\ldots$?

5. Mathematical use of language sometimes differs from colloquial usage. In each of the following statements, record whether you think that they are true or false. Keep them for comparison when you read chapter 6.
 (a) All of the numbers 2, 5, 17, 53, 97 are prime.
 (b) Each of the numbers 2, 5, 17, 53, 97 is prime.
 (c) Some of the numbers 2, 5, 17, 53, 97 are prime.
 (d) Some of the numbers 2, 5, 17, 53, 97 are even.
 (e) All of the numbers 2, 5, 17, 53, 97 are even.
 (f) Some of the numbers 2, 5, 17, 53, 97 are odd.

6. 'If pigs had wings, they'd fly.'
Is this a logical deduction?

7. 'The set of natural numbers 1, 2, 3, 4, 5,... etc. is infinite.' Give an explanation of what the word 'infinite' means in this context.

8. A formal definition of the number 4 might be given in the following terms.
 First note that a set is specified by writing its elements between curly brackets $\{\,,\}$ and that the set with no elements is denoted by $\varnothing$. Then we define

$$4 = \{\varnothing, \{\varnothing\}, \{\varnothing, \{\varnothing\}\}, \{\varnothing, \{\varnothing\}, \{\varnothing, \{\varnothing\}\}\}\}.$$

Can you understand this definition? Do you think that it is suitable for a beginner?

9. Which, in your opinion, is the most likely explanation for the equality

$$(-1)\times(-1) = +1?$$

 (a) A scientific truth discovered by experience,
 (b) A definition formulated by mathematicians as being the only sensible way to make arithmetic work,
 (c) A logical deduction from suitable axioms,

(d) Some other explanation.
Give reasons for your choice and retain your comments for later consideration.

10. In multiplying two numbers together, the order does not matter, $xy = yx$. Can you justify this result
   (a) When x, y are both whole numbers,
   (b) when x, y are any real numbers,
   (c) for any numbers whatever?

2

Number systems

THE reader will have built up a coherent understanding of the arithmetic of the various number systems: counting numbers, negative numbers, and so on. But he may not have subjected the processes of arithmetic to close logical scrutiny. Later in the book it is our intention to place these number systems in a precise axiomatic setting. In this chapter we shall give a brief review of the way that the reader may have developed his ideas on these systems. Although constant use of the ideas will have smoothed out many of the difficulties which were encountered when the concepts were being formed, these difficulties tend to reappear in the formal treatment and have to be dealt with again. It is therefore very much worth spending a little time to recall the development, before we plunge into the formalities. The experienced reader may feel tempted to skip this chapter because of the very simple level of the discussion. Please don't. Every adult's ideas have been built up from simple beginnings as a child. If we are trying to understand the foundations of mathematics, it is important to be aware of the genesis of our own mathematical thought-processes.

The Natural numbers

The natural numbers are the familiar counting numbers 1, 2, 3, 4, 5, ... Young children learn the names of these, and the order in which they come, by rote. Contact with adults leads the children to an awareness of the meaning that adults attach to phrases such as 'two sweets', 'four marbles'. Use of the word 'zero' and the concept 'no sweets' is a little harder and follows shortly afterwards.

To count a collection of objects, we just point to them in turn whilst reciting 'one, two, three, . . . ' until we have pointed to all of the objects, once each.

Next we learn the arithmetic of natural numbers, starting with addition. At this stage the basic 'laws' of addition (which we can express algebraically as the commutative law $a + b = b + a$, and the associative law $a + (b + c) = (a + b) + c$ may or may not be 'obvious', depending on the approach used. If addition is introduced in terms of combining the actual collections of objects and then counting the result, then these two laws depend only on the tacit assumption that rearranging the collection does not alter the number of things in it. Similarly one modern approach using coloured rods whose lengths represent the numbers (which are added by placing them end to end) makes commutativity and associativity 'non-concepts', so obvious that it is almost confusing to have them pointed out. However, if a child is taught addition by 'counting on', the story is quite different. To calculate $3 + 4$ he starts at 3 and counts on four more places: 4, 5, 6, 7. To calculate $4 + 3$ he starts at 4 and counts on three places: 5, 6, 7. That the two processes yield the same answer is now much more mysterious. In fact children taught this way often have difficulty doing a calculation such as $1 + 17$, but find $17 + 1$ trivial!

Next we come to the concept of place-value. The number 33 involves two threes, but they don't mean the same thing. It must be emphasized that this is purely a matter of notation, and has nothing to do with the numbers *themselves*. But it is a highly useful and important notation. It can represent (in principle) arbitrarily large numbers, and is very well adapted to calculation. However, a precise mathematical description of the general processes of arithmetic in Hindu–Arabic place notation is quite complicated (which is why children take so long to learn them all) and not well adapted to, say, a proof of the commutative law. Sometimes a more primitive system has some advantages. For instance, the ancient Egyptians used the symbol | to represent 1, a hoop ∩ to represent 10, the end of a scroll ☉ for 100, with other symbols for 1000, etc. A number was written by repeating these symbols: thus 247 would have been written

$$☉☉∩ ∩ ∩ ∩ \ ||||||| $$

Adding in Egyptian is easy: all one does is to put the symbols together. Now the commutative and associative laws are obvious again. But the notation is less suited to computation. To recover place-notation from Egyptian all we need do is add some 'carrying rules', such as $||||||||| = ∩$, and to insist that we never use any particular symbol more than nine times.

Before proceeding, we shall introduce a small amount of notation. We write $\mathcal{N}$ for the set of all natural numbers. The symbol ∈ will mean 'is an element of' or 'belongs to'. So the symbols

$$2 \in \mathcal{N}$$

are read as '2 belongs to the set of natural numbers', or in more usual language, '2 is a natural number'.

Fractions

Fractions are introduced to allow division. It is easy to divide 12 into 3 parts: $12 = 4 + 4 + 4$. It is not possible to divide, say, 11 into 3 equal parts if we insist that these parts be natural numbers. Hence we are led to define fractions m/n where m, $n \in \mathcal{N}$ and $n \neq 0$. Operations of addition and multiplication on the set $\mathcal{F}$ of fractions can be defined: algebraically expressed they are

$$\frac{m}{n} + \frac{p}{q} = \frac{mq + np}{nq}$$

$$\frac{m}{n} \cdot \frac{p}{q} = \frac{mp}{nq}$$

The notation has one disadvantage in that two apparently different fractions may have the same value. For example, $2/3 = 4/6$. This is overcome by allowing the 'cancellation' of common factors, reducing the fraction to 'lowest terms'.

The integers

What fractions do for division, the integers do for subtraction. A subtraction sum like $2 - 7 = ?$ cannot be answered in $\mathcal{N}$. We must introduce negative numbers. Children are often introduced to negative numbers in terms of a 'number line'. A straight line is drawn and equally spaced points are marked on it. One is called 0, then natural numbers 1, 2, 3, ... are marked successively to the right, negative numbers -1, -2, -3, ... to the left.

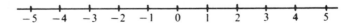

This gives an extended number system called the 'integers'. An integer is either a natural number n, or a symbol $-n$ where n is a natural number, or 0. We use $\mathcal{Z}$ to denote the set of integers. (Z is the initial letter of 'Zahlen', the German for integers.)

The arithmetic of $\mathcal{Z}$ is moderately complicated to describe. The device of thinking of negative numbers as debts is often used. Multiplication of negative numbers has to get across the point that 'minus times minus makes plus'.

We shall make premature use of another piece of set-theoretic notation. The symbol $\subseteq$ will mean 'is a subset of'. We then have

$$\mathcal{N} \subseteq \mathcal{Z}$$

since every natural number is also an integer. Similarly we have

$$\mathcal{N} \subseteq \mathcal{F}.$$

There is a species of mathematician that denotes the 'positive integers' by $+1, +2, +3$, and so on, by analogy with the 'negative integers' $-1, -2, -3, \ldots$. For precision, they distinguish the natural numbers, as used for counting, from the positive integers. It is unfair to pillory such people for making what may seem a needless distinction, because it is sometimes a convenient one on technical grounds. There was once a Greek scholar who maintained that the *Odyssey* was not written by Homer, but by another man *of the same name*. If you have good reason to distinguish between the two (bearing in mind that almost nothing is known about Homer in any case) this kind of thing is perfectly justified; but for most ordinary purposes the distinction is quite meaningless. The same thing should be said about the natural numbers and the positive integers.

Rational numbers

The system $\mathcal{Z}$ is designed to allow subtraction in all cases; the system $\mathcal{F}$ to allow division (except by zero). However, in neither system are *both* operations always possible. To get both working at once we move into the system of 'rational numbers' denoted by $\mathcal{Q}$ (a script Q, for 'quotients'). This is obtained from $\mathcal{F}$ by introducing 'negative fractions' in much the same way that we obtained $\mathcal{Z}$ from $\mathcal{N}$.

We can still represent $\mathcal{Q}$ by points on a 'number line', by marking fractions at suitably spaced intervals between the integers, with negative ones to the left of 0 and positive to the right. For example, $4/3$ is marked one third of the way between 1 and 2, like this:

The rules for adding or multiplying rational numbers are the same as for fractions, but now m, n, p, q are allowed to be integers, rather than just natural numbers.

Both $\mathcal{X}$ and $\mathcal{F}$ are subsets of $\mathcal{Q}$. We can summarize the relations between the four number systems so far encountered by the diagram

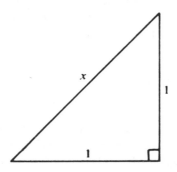

Real numbers

One may use numbers to measure lengths, or other physical quantities. However the Greeks discovered that there exist lines whose lengths, in theory, cannot be measured exactly by a *rational* number. They were magnificent geometers, and one of their simple but profound results was Pythagoras's theorem. Applied to a right-angled triangle whose two shorter sides have lengths 1, this implies that the hypotenuse has length x, where $x^2 = 1^2 + 1^2 = 2$. Now: there is no

rational number m/n such that $(m/n)^2 = 2$. To see this we shall use the result that any natural number can be factorized uniquely into primes. For instance, we can write

$$360 = 2 \times 2 \times 2 \times 3 \times 3 \times 5$$

or

$$360 = 5 \times 2 \times 3 \times 2 \times 3 \times 2$$

but however we write the factors we will always have one 5, two 3's and three 2's. Using index notation we write

$$360 = 2^3 . 3^2 . 5.$$

We shall prove this unique factorization theorem formally in chapter 8 but for the moment we shall assume it without further proof.

If we factorize any natural number into primes and then square, each prime will now occur an *even* number of times. For instance,

$$360^2 = (2^3 . 3^2 . 5)^2 = 2^6 . 3^4 . 5^2,$$

and the indices 6, 4, 2 are all even. A general proof is not hard to find.

Now take any rational number m/n and square it. (Since m/n has the same square as $-m/n$, we may assume m and n positive.) Factorize m^2 and n^2 and cancel factors top and bottom if possible. Whenever a prime p cancels, then since all primes occur to even powers it follows that p^2 cancels. Hence after cancellation, all primes still occur to even powers. But $(m/n)^2$ is supposed to equal 2, which has one prime (namely 2) which only occurs once (which is an odd power).

It follows that no rational number can have square 2, and hence that the hypotenuse of the given triangle does not have rational length.

By using a little more algebraic symbolism one can tidy up this proof and present it as a formal argument; but the above is all that we really need. The same argument shows that numbers like 3, 3/4, or 5/7 do not have *rational* square roots.

If we want to talk of lengths like $\sqrt{2}$ we must further enlarge our number system. Not only do we need rational numbers, but 'irrational' ones as well.

Using Hindu–Arabic notation this can be done if we introduce decimal expansions. We construct a right-angled triangle with sides of unit length, and using drawing instruments transfer the length of its hypotenuse to the number line. We then obtain a specific point on the number line which we call $\sqrt{2}$. It lies between 1 and 2 and, on

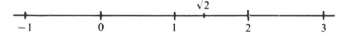

subdividing the unit length from 1 to 2 into ten equal parts, we find that $\sqrt{2}$ lies between 1·4 and 1·5. By further subdividing the distance

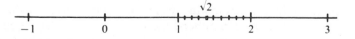

between 1·4 and 1·5 into ten equal parts we might hope to obtain a better approximation to √2. Already in a practical situation we are reaching the limit of accuracy in drawing. We might imagine that in an accurate diagram we can look sufficiently close, or magnify the picture to give the next decimal place. If we were to look at an actual picture under a magnifying glass, not only would the lengths be magnified, but

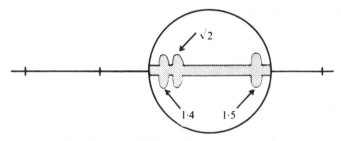

so would the thickness of the lines in the drawing. This would not be a very satisfactory way to obtain a better estimate for √2.

Practical drawing is in fact extremely limited in accuracy. A fine drawing pen marks a line 0·1 millimetres thick. Even if we use a line 1 metre long as a unit length, since 0·1 mm. = 0·0001 metres, we could not hope to be accurate to more than four decimal places. Using much larger paper and more refined instruments gives surprisingly little increase in accuracy in terms of the number of decimal places we can find. A light year is approximately $9·5 \times 10^{15}$ metres. As an extreme case, suppose we consider a unit length 10^{18} metres long. If a light ray started out at one end at the same time that a baby was born at the other, the baby would have to live to be over 100 years old before he would see the light ray. At the lower extreme of vision, the wavelength of red light is approximately 7×10^{-7} metres, so a length of 10^{-7} metres is smaller than the wavelength of visible light. Hence an ordinary optical microscope cannot distinguish points which are 10^{-7} metres apart. On a line where the unit length is 10^{18} metres we cannot distinguish numbers which are less than $10^{-7}/10^{18} = 10^{-25}$ apart. This means that we cannot achieve an accuracy of 25 decimal places by a drawing. Even this is a gross exaggeration in practice, where three decimal places is the best we can really hope for.

Inaccurate arithmetic in practical drawing

The inherent inaccuracy in practice leads to problems in arithmetic. If we add two inaccurate numbers, the errors also add. If we cannot distinguish errors less than some amount e, then we cannot tell the difference, in practice, between a and $a + \frac{3}{4}e$ and between b and $b + \frac{3}{4}e$. But adding, we *can* distinguish between $a + b$ and $a + b + \frac{3}{2}e$. When we come to multiplication, errors can increase even more dramatically. We cannot hope to get answers to the same degree of accuracy as the numbers used in the calculation.

If we do our arithmetic and calculate all answers correct to a certain number of decimal places, the errors involved lead to some disturbing results. Suppose, for example, that we work to two decimal places ('rounding up' if the third place is 5 or more and down if it is less). Given two real numbers a and b, we denote their product correct to two decimal places by $a \otimes b$. For example, $3{\cdot}05 \otimes 4{\cdot}26 = 12{\cdot}99$ because $3{\cdot}05 \times 4{\cdot}26 = 12{\cdot}993$. Using this law of multiplication we find that

$$(1{\cdot}01 \otimes 0{\cdot}5) \otimes 10 \neq 1{\cdot}01 \otimes (0{\cdot}5 \otimes 10).$$

The left-hand side reduces to $0{\cdot}51 \otimes 10 = 5{\cdot}1$, whilst the right-hand side becomes $1{\cdot}01 \otimes 5 = 5{\cdot}05$. This is by no means an isolated example, and it shows that the associative law does not hold for $\otimes$.

If we further define $a \oplus b$ to be the sum, correct to two decimal places, we will find other laws which do not hold, including the distributive law $a \otimes (b \oplus c) \overset{?}{=} (a \otimes b) \oplus (a \otimes c)$.

A theoretical model of the real line

We have just seen that if our measurement of numbers is not precise, then some of the laws of arithmetic break down. To avoid this we must make our notion of real number exact.

Let us suppose that we are given a real number x on a theoretical real line, and try to express it as a decimal expansion. As a starting

point we see that x lies between two integers. In the above example x is between 2 and 3, so x is 'two point something'. Next we divide the interval between 2 and 3 into ten equal parts. Again, x lies in some sub-interval. In the picture, x lies between $2{\cdot}4$ and $2{\cdot}5$, so x is '$2{\cdot}4$

something'. To obtain a better idea, we divide the interval between 2·4 and 2·5 into ten equal parts, and repeat the process to find the next figure in the decimal expansion. Already, in a practical situation, we are reaching the limit of accuracy in drawing.

For our theoretical picture we must imagine that we can look sufficiently closely, or magnify the picture, to read off the next decimal place. If we looked at an actual picture under a magnifying glass, not only would the lengths be magnified, but so would the thickness of the lines. This is not very satisfactory for getting a better estimate. We

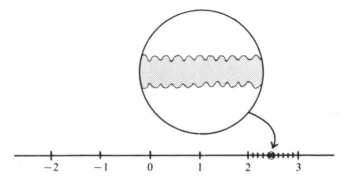

must, in the theoretical case, assume that the lines have no thickness, so that they are not made wider when the picture is magnified. We can represent this as a practical picture by drawing the magnified lines with the same drawing implements as before, and making them as fine as possible. In this case x lies between 2·43 and 2·44, so x is '2·44

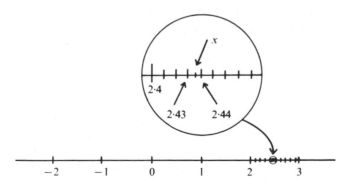

something'. Using this method we can, in theory, represent any real number as a decimal expansion to as many figures as we require. Two numbers will be different if, by calculating sufficiently many terms, we eventually obtain different answers for some decimal place.

We can express this theoretical method as follows in more mathematical terms.

(i) Given a real number x, find an integer a_0 such that

$$a_0 \le x < a_0 + 1.$$

(ii) Find a whole number a_1 between 0 and 9 inclusive such that

$$a_0 + \frac{a_1}{10} \le x < a_0 + \frac{a_1 + 1}{10}.$$

(iii) After finding $a_0, a_1, \ldots, a_{n-1}$, where $a_1, \ldots, a_{n-1}$ are integers between 0 and 9 inclusive, find the integer a_n between 0 and 9 inclusive for which

$$a_0 + \frac{a_1}{10} + \ldots + \frac{a_n}{10^n} \le x < a_0 + \frac{a_1}{10} + \ldots + \frac{a_n + 1}{10^n}.$$

This gives an inductive process which at the nth stage determines x to n decimal places:

$$a_0 \cdot a_1 a_2 \ldots a_n \le x < a_0 \cdot a_1 a_2 \ldots a_n + 1/10^n.$$

The theoretically exact representation of the number x requires a decimal expansion

$$a_0 \cdot a_1 a_2 a_3 a_4 a_5 a_6 \ldots$$

which goes on forever. (Of course, if all a_n from some point on are zero, we omit them in normal notation: instead of $1066 \cdot 31700000000 \ldots$ we write $1066 \cdot 317$.) An infinite decimal is called a *real number*. The set of all real numbers is denoted by $\mathcal{R}$.

In most practical situations we will need only a few decimal places. The observations of the first section of this chapter show that 25 decimal places are sufficient for all ratios of lengths within the human capacity for visual perception. Of course this is a gross overestimate, and two or three places are usually quite good enough for the practical man.

Different decimal expansions for different numbers

If we expand a number x as above in an endless decimal, we say that $a_0 \cdot a_1 a_2 \ldots a_n$ is the expansion of x *to the first n decimal places* (without 'rounding up'). If two real numbers x and y have the same decimal expansion to n places then

$$a_0 \cdot a_1 a_2 \dots a_n \leqslant x < a_0 \cdot a_1 a_2 \dots a_n + 1/10^n,$$

$$a_0 \cdot a_1 a_2 \dots a_n \leqslant y < a_0 \cdot a_1 a_2 \dots a_n + 1/10^n.$$

The second line of inequalities can be rewritten as

$$-a_0 \cdot a_1 \dots a_n - 1/10^n < -y \leqslant -a_0 \cdot a_1 \dots a_n.$$

Adding this to the first line we obtain

$$-1/10^n < x - y < 1/10^n.$$

In other words, if two real numbers have the same decimal expansion to n places they differ by at most $1/10^n$.

If x and y are *different* numbers on our line and we wish to distinguish between them, all we need do is find n such that $1/10^n$ is less than their difference: then their expansion to n places will differ. This again exposes the deficiences of practical drawing, where x and y might be too close to distinguish. In our theoretical concept of the real line, this distinction must always be possible. It is so important that it is worth giving it a name. The great Greek mathematician Archimedes stated a property which is equivalent to what we want, so we shall name our condition after him:

Archimedes' Condition: Given a positive real number ε, there exists a positive integer n such that $1/10^n < \varepsilon$.

Rationals and Irrationals

As we have seen, the real number $\sqrt{2}$ is irrational: so are many others. It is not always easy to prove a given number irrational. (It's moderately easy for e, less so for π, and there are many interesting numbers which mathematicians have been convinced for centuries are irrational, but have never proved them to be.) But just the fact that $\sqrt{2}$ is irrational implies that between any two rational numbers there exist irrational numbers. First we need:

LEMMA 1.   If m/n and r/s are rational, with $r/s \neq 0$, then $m/n + (r/s)\sqrt{2}$ is irrational.

PROOF.   Suppose that $m/n + (r/s)\sqrt{2}$ is rational, equal to p/q where p, q are integers. Then we may solve for $\sqrt{2}$ to obtain

$$\sqrt{2} = (pn - mq)s/qnr$$

which is rational, contrary to the irrationality of $\sqrt{2}$.   □

PROPOSITION 2.   Between any two distinct rational numbers there exists an irrational number.

PROOF.   Let the rational numbers be m/n and r/s, where $m/n < r/s$. Then

$$m/n < m/n + \frac{\sqrt{2}}{2}(r/s - m/n) < r/s$$

(because $\sqrt{2}/2 < 1$), and the number in the middle is irrational by the lemma.   □

There is a corresponding result with 'rational' and 'irrational' interchanged:

PROPOSITION 3.   Between any two distinct irrational numbers there exists a rational number.

PROOF.   Let the irrational numbers be a, b with $a < b$. Consider their decimal expansions, and let the nth decimal place be the first in which they differ. Then

$$a = a_0 \cdot a_1 \ldots a_{n-1} a_n \ldots,$$

$$b = a_0 \cdot a_1 \ldots a_{n-1} b_n \ldots,$$

where $a_n \neq b_n$. Let $x = a_0 \cdot a_1 \ldots a_{n-1} b_n$. Then x is rational, and we must have $a < x < b$. (Obviously $a < x \leq b$, but since b is irrational, $x \neq b$.)   □

In fact the exercises at the end of this chapter show that the rational and irrational numbers are mixed up in a very complicated way. One should *not* make the mistake of thinking that they 'alternate' along the real line.

The rational numbers may be characterized as those whose decimal expansions repeat at regular intervals (though we shall omit the proof). To be precise, say that a decimal is *repeating* if, from some point on, a fixed sequence of digits repeats indefinitely. For example, $1 \cdot 5432174174174174 \ldots$ is a repeating decimal. We shall write it as $1 \cdot 5432 \dot{1} 7 \dot{4}$, with dots over the end digits of the block that repeats.

The need for real numbers

We have seen that the belief of the Greeks that all numbers were rational (enshrined in the mystic philosophy of the cult of Pythagoreans) led them to an impasse. Viewing the real numbers as infinite decimals helps to overcome this mental block, because it makes it clear that rational numbers, whose expansions repeat, do not exhaust the possibilities.

However, we have also seen that for practical purposes we do not need infinite decimals, nor even very long finite ones. Why go to all the trouble? One reason we have already noted: the arithmetic of decimals of limited length fails to obey the familiar laws which integers or rational numbers obey. A perhaps more serious reason arises in analysis.

Consider the function given by

$$f(x) = x^2 - 2 \qquad (x \in \mathcal{R}).$$

This is negative at $x = 1$, positive at $x = 2$. In between it is zero at $x = \sqrt{2}$. However, if we restrict x to take only rational values, the function

$$f(x) = x^2 - 2 \qquad (x \in \mathcal{Q})$$

is also negative at $x = 1$, positive at $x = 2$, but is not zero at any rational x in between, because $x^2 = 2$ has no *rational* solution. This is a

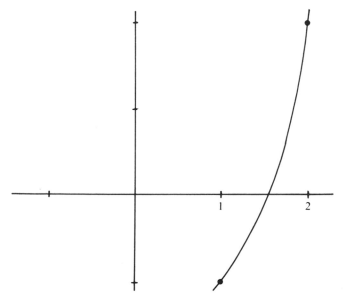

nuisance. A fundamental theorem in analysis asserts that if a continuous function is negative at one point and positive at another, then it must be zero in between. This is true for functions over the real numbers, but not for functions over the rationals. A civilization such as that of the ancient Greeks, with no satisfactory method for handling irrational numbers, cannot build a theory of limits, or invent calculus.

Arithmetic of decimals

The idea of infinite decimals representing real numbers is a useful one, but it is not well suited to numerical manipulations, nor to theoretical investigations beyond an elementary level. Thus we add two finite decimals by starting at the right-hand end: but infinite decimals do not have right-hand ends and there is nowhere to start.

We can instead start at the left-hand end, adding the first decimal places, then the first two, then the first three, and so on. To see what happens, try adding $2/3 = 0 \cdot \dot{6}$ and $2/7 = 0 \cdot \dot{2}8571\dot{4}$ in this way.

$$
\begin{array}{llll}
\cdot 6 & + \cdot 2 & = \cdot 8 \\
\cdot 66 & + \cdot 28 & = \cdot 94 \\
\cdot 666 & + \cdot 285 & = \cdot 951 \\
\cdot 6666 & + \cdot 2857 & = \cdot 9523 \\
\cdot 66666 & + \cdot 28571 & = \cdot 95237 \\
\cdot 666666 & + \cdot 285714 & = \cdot 952380.
\end{array}
$$

The actual answer is $2/3 + 2/7 = 20/21 = 0 \cdot \dot{9}5238\dot{0}$. Notice that adding the first decimal places does not give the answer to one decimal place, nor does adding the first two places give the first two places of the answer. This is precisely because of the possibility of 'carried' digits from later places affecting earlier ones.

However, in this example, successive terms increase and get closer and closer to the actual answer. The sequence of numbers $\cdot 8$, $\cdot 94$, $\cdot 951$, $\cdot 9523, \dots$ is an increasing sequence of real numbers, and it 'tends to' $20/21$ in the sense that the error can be made as small as we please by calculating enough decimal places.

In the next few sections we shall examine in detail the ideas required to make this concept precise. For theoretical purposes it is often easier to use increasing sequences (of approximations to a real number) rather than decimal expansions.

Sequences

A *sequence* of real numbers can be thought of as an endless list

$$ a_1, a_2, a_3, a_4, \dots, a_n, \dots $$

where each a_n is a real number. (Using set theory we shall give a more formal definition in chapter 5). For example,

(1) The sequence of squares: 1, 4, 9, 16, ... where $a_n = n^2$.

(2) The sequence of decimal approximations to $\sqrt{2}$: 1·4, 1·41, 1·414, ... where $a_n = \sqrt{2}$ to n places.

(3) The sequence 1, $1\frac{1}{2}$, $1\frac{5}{6}$, ... where $a_n = 1 + \frac{1}{2} + \frac{1}{3} + \ldots + \frac{1}{n}$.

(4) The sequence 3, 1, 4, 1, 5, 9, ... where a_n = the nth digit in the decimal expansion of π.

A shorthand notation for the sequence $a_1, a_2, \ldots$ which we shall often use is

$$(a_n)$$

where the nth term is placed in round brackets. Thus example (1) could be written (n^2).

Notice how general the concept of a sequence is. We can consider *any* endless list of numbers. It is not necessary that the nth term be defined by a 'nice formula', as long as we know what each a_n is supposed to be.

Sequences can be added, subtracted, or multiplied. It is necessary to define what we mean by this: the simplest way is to perform the operations on each pair of terms in corresponding positions. In other words, to add the sequences

$$a_1, a_2, \ldots$$

and

$$b_1, b_2, \ldots$$

means to form the sequence

$$a_1 + b_1, a_2 + b_2, \ldots.$$

For example, if $a_n = n^2$ and $b_n = 1 + \frac{1}{2} + \ldots + \frac{1}{n}$, then the nth term of $(a_n) + (b_n)$ is

$$n^2 + 1 + \frac{1}{2} + \ldots + \frac{1}{n}.$$

Since the nth term of the sequence $(a_n) + (b_n)$ is $a_n + b_n$, we can express the rule for addition as

$$(a_n) + (b_n) = (a_n + b_n).$$

Similarly the rules for subtraction and multiplication are given by

$$(a_n) - (b_n) = (a_n - b_n),$$
$$(a_n)(b_n) = (a_n b_n).$$

In the case of division we put

$$(a_n)/(b_n) = (a_n/b_n),$$

but note that this division can be carried out only when *all* terms b_n are non-zero.

EXAMPLE.   If $a_n = \sqrt{2}$ to n decimal places, and $b_n =$ the nth decimal place in π, then the first few terms of $(a_n)(b_n)$ are

$$1{\cdot}4 \times 3 = 4{\cdot}2$$
$$1{\cdot}41 \times 1 = 1{\cdot}41$$
$$1{\cdot}414 \times 4 = 5{\cdot}656$$
$$1{\cdot}4142 \times 1 = 1{\cdot}4142.$$

If you were given the sequence $4{\cdot}2$, $1{\cdot}41$, $5{\cdot}656$, $1{\cdot}4142$, could you have guessed the rule for the nth term? This drives home the way in which we must know in principle how to calculate *all* the terms of the sequence, in order to specify it. In general it is not enough to write down the first few terms and a few dots. The sequence 3, 1, 4, 1, 5, 9, ... certainly looks as if it consists of the digits of π. However, it might just as well be the sequence of digits of the number 355/113, which starts off the same way. This is why, in example (4), we specify the general rule for finding the nth term.

Nevertheless, you will often find mathematicians writing things like 2, 4, 8, 16, 32, ... and expecting you to infer that the nth term is 2^n. One aspect of learning mathematics is to understand how mathematicians actually work, and what their idiosyncracies are: you should be prepared to accept slight differences in notation provided that the idea is clear from the context.

Order properties and the modulus

We digress from our train of thought to introduce an important concept which will be of great use to us in a moment. If x is a real number we define the *modulus* or *absolute value* of x to be

$$|x| = \begin{cases} x & \text{if } x \geq 0, \\ -x & \text{if } x < 0. \end{cases}$$

The graph of $|x|$ against x looks like:

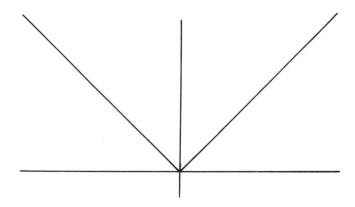

The value of $|x|$ tells us how large or small x is, ignoring whether it is positive or negative. Perhaps the most useful fact about the modulus is the *triangle inequality*, so called because its generalization to complex numbers expresses the fact that each side of a triangle is shorter than the other two put together. It is:

PROPOSITION 4.   (Triangle inequality). If x and y are real numbers, then

$$|x + y| \leq |x| + |y|.$$

PROOF.   The idea is that $|x + y|$ says how far from the origin $x + y$ is, and this is at most the sum of the distances $|x|$ and $|y|$ of x and y from the origin, being less if x and y have opposite sign. (Draw some pictures to check this.) The easiest way to prove it is to divide into cases, according to the signs and relative sizes of x and y.

(i) $x \geq 0$, $y \geq 0$. Then $x + y \geq 0$, so we have

$$|x + y| = x + y = |x| + |y|.$$

(ii) $x \geq 0$, $y < 0$. If $x + y \geq 0$ then

$$|x + y| = x + y < x - y = |x| + |y|.$$

On the other hand, if $x + y < 0$ then

$$|x + y| = -(x + y) = -x - y < |x| + |y|.$$

(iii) $x < 0$, $y \geq 0$ follows as in case (i) with x and y interchanged.

(iv) $x < 0$, $y < 0$. Then $x + y < 0$, so we have

$$|x + y| = -x - y = |x| + |y|.$$

This completes the proof in all cases.   □

One should be on the lookout for variations on this theme, such as

$$|x - y| + |y - z| \geq |x - z|,$$

which follows since $x - z = (x - y) + (y - z)$, so that $|x - z| \leq |x - y| + |y - z|$.

The modulus is most useful for expressing certain inequalities succinctly. For example,

$$a - \varepsilon < x < a + \varepsilon$$

can be written

$$-\varepsilon < x - a < \varepsilon,$$

which translates into

$$|x - a| < \varepsilon.$$

Convergence

Now we are ready to consider the general notion of representing a real number as a 'limit' of a sequence, rather than just being a particular decimal expansion. As an exercise, the reader should mark, to as large a scale and as accurately as possible, the numbers 1·4, 1·41, 1·414, 1·4142, $\sqrt{2}$, on the interval between 1 and 2.

Notice that the numbers 1·4, 1·41, 1·414, 1·4142, get closer and closer together, until they become indistinguishable from each other and from $\sqrt{2}$, up to the accuracy of the drawing. By drawing a more accurate picture we must go further along the sequence of decimal approximations to $\sqrt{2}$ before this happens. Thus if we work to an accuracy of 10^{-8}, then from the eighth term onwards all points of the sequence are indistinguishable from $\sqrt{2}$.

This observation motivates the theoretical concept of *convergence*. Let ε be any positive real number (ε is the Greek letter for 'e' and may be thought of as the initial letter of 'error'). For practical convergence of a sequence (a_n) to a limit l, if we are working to an accuracy ε, we require there to be some natural number N such that the difference between a_n and l has size less than ε when $n > N$. In other words, $|a_n - l| < \varepsilon$. In the following diagram we cannot distinguish points less than ε apart: in this case $N = 7$ and a_n is indistinguishable from l when $n > 7$.

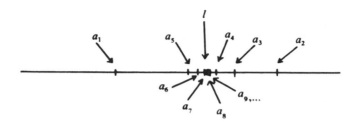

For theoretical convergence we ask that a similar phenomenon occur for *all* positive ε. This is on the explicit understanding that smaller ε's may require larger N's. In this sense, N is allowed to depend on ε. Thus we reach:

DEFINITION.   A sequence (a_n) of real numbers tends to a limit l if, given any $\varepsilon > 0$, there is a natural number N such that

$$|a_n - l| < \varepsilon$$

for all $n > N$.

Mathematicians use various pieces of shorthand notation to express this concept. To say 'the sequence (a_n) tends to the limit l' one writes

$$\lim_{n \to \infty} a_n = l,$$

or

$$a_n \to l \text{ as } n \to \infty.$$

The compound symbol '$n \to \infty$' is meant to remind us that we are interested in the behaviour of a_n as n becomes large (namely $n > N$). The symbol '∞' on its own has not been given a meaning. It is *not* a number which is 'arbitrarily large', or anything of that nature: it isn't even a number. In fact it is often better to forget it entirely, and just write

$$\lim a_n = l,$$

but bearing in mind that mathematicians use other kinds of limits elsewhere, which might be confused with this one.

EXAMPLE.   The sequence $1 \cdot 1$, $1 \cdot 01$, $1 \cdot 001$, $1 \cdot 0001$, ..., for which $a_n = 1 + 10^{-n}$ tends to the limit 1 as $n \to \infty$. For, given $\varepsilon > 0$, we have to make

$$|1 + 10^{-n} - 1| < \varepsilon \quad \text{for } n > N,$$

by finding a suitable N. But this follows by Archimedes' condition: if we find N to make $10^{-N} < \varepsilon$, then for all $n > N$ we have $10^{-n} < 10^{-N} < \varepsilon$. (If we have the theory of logarithms available, all we have to do is take $N > \log_{10}(1/\varepsilon)$.)

· A sequence (a_n) which tends to a limit l is called *convergent*. If no limit exists, it is *divergent*.

It should be remarked that a convergent sequence can tend to only one limit. For suppose $a_n \to l$ and $a_n \to m$, where $l \neq m$. We take $\varepsilon = \frac{1}{2}|l - m|$. For large enough n, we have

$$|a_n - l| < \varepsilon,$$
$$|a_n - m| < \varepsilon.$$

So, from the triangle inequality, $|l - m| < 2\varepsilon = |l - m|$, which is not the case.

In other words, if all the terms a_n must eventually be very close to l they cannot also be very close to m, because this requires them to be in two different places at the same time.

Completeness

A sequence (a_n) is *increasing* if each $a_n \leqslant a_{n+1}$, so that

$$a_1 \leqslant a_2 \leqslant a_3 \leqslant \ldots .$$

Suppose that (a_n) is an increasing sequence. Either the terms a_n increase without limit, eventually becoming as large as we please, or else there must be some real number k such that $a_n \leqslant k$ for *all* n. An example of a sequence of the first type is $1, 4, 9, 16, 25, \ldots$; one of the latter type is the sequence of decimal approximations to e: $2 \cdot 7$, $2 \cdot 71$, $2 \cdot 718$, $2 \cdot 7182$, ..., every term of which is less than 3.

If there exists a real number k such that $a_n \leqslant k$ for all n we say that (a_n) is *bounded*.

If we draw the points of a bounded increasing sequence on a part of the real line we need only draw the interval between a_1 and k, since all the other points lie inside this. So a typical picture is

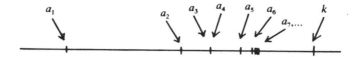

Intuitively it is clear that the terms become increasingly squashed together, and tend to some limit $l \leqslant k$. This intuition is correct if we consider sequences of real numbers and real limits, but it is wrong for sequences of rational numbers and *rational* limits. In fact the sequence of decimal approximations to $\sqrt{2}$ is an increasing sequence of rational numbers with no rational number as limit.

The fact that every bounded sequence of real numbers tends to a real number as limit as known as the *completeness* property of the real numbers. (The force of the name is that the rational numbers are 'incomplete' because numbers like $\sqrt{2}$ are 'missing').

We can make the completeness property of the reals very plausible in terms of our ideas about decimals. Let (a_n) be an increasing sequence of real numbers, with $a_n \leqslant k$ for all k.

Now the set of integers between $a_1 - 1$ and k is finite, so there is an integer b_0 which is the largest integer for which some term a_{n_0} of the sequence is $\geqslant b_0$.

Then all terms a_n are less than $b_0 + 1$.

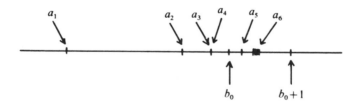

We subdivide the interval from b_0 to $b_0 + 1$ into ten parts, and find b_1 so that some term $a_{n_1} \geqslant b_0 + b_1/10$, but no term $a_n \geqslant b_0 + (b_1 + 1)/10$. Continuing in this way we get a sequence of decimals

$$b_0, \; b_0 \cdot b_1, \; b_0 \cdot b_1 b_2, \ldots$$

such that for $n > n_r$ we have a_n lying between $b_0 \cdot b_1 \ldots b_r$ and $b_0 \cdot b_1 \ldots b_r + 1/10^r$. Then the real number

$$l = b_0 \cdot b_1 b_2 \ldots$$

has the property that $|a_n - l| < 1/10^r$ for all $n > n_r$. Hence $a_n \to l$ as $n \to \infty$.

It is easy to check that this l is less than or equal to k.

Decreasing sequences

There is no need to be obsessed with increasing sequences. A sequence (a_n) is *decreasing* if $a_n \geqslant a_{n+1}$ for all n. If it satisfies $a_n \geqslant k$ for all n then k is a *lower bound* and the sequence is *bounded below*. (To avoid ambiguity with increasing sequences we can now say *bounded above* instead of 'bounded'.) There is a similar theorem concerning decreasing sequences, but instead of copying out the proof again and changing the inequalities we use a trick. If (a_n) is decreasing, then $(-a_n)$ is increasing. If $a_n \geqslant k$ for all n then $-a_n \leqslant -k$ for all n, so $(-a_n)$ is bounded above, hence tends to a limit l. It follows easily that $a_n \to -l$. Hence any decreasing sequence of real numbers bounded below by k tends to a limit $-l \geqslant k$.

Different decimal expansions for the same real number

Previously we expanded a real number as an infinite decimal, $x = a_0 \cdot a_1 a_2 \ldots$, by using the inequalities

$$a_0 + \frac{a_1}{10} + \ldots + \frac{a_n}{10^n} \leqslant x < a_0 + \frac{a_1}{10} + \ldots + \frac{a_n + 1}{10^n},$$

where a_0 is an integer and a_n is an integer from 0 to 9 for $n \geqslant 1$. This condition can be written

$$a_0 \cdot a_1 a_2 \ldots a_n \leqslant x < a_0 \cdot a_1 a_2 \ldots a_n + 1/10^n. \tag{*}$$

This, used successively for $n = 1, 2, 3, \ldots$ gives a unique decimal expansion for any real number, and different real numbers have different decimal expansions. However, this is not quite the whole story since certain decimal expansions do not occur when we use condition (*). For example the expansion $0 \cdot 999999 \ldots$ where $a_0 = 0$ and $a_n = 9$ for all $n \geqslant 1$, does not occur.

Let us see why this is true. Suppose there were a real number x with decimal expansion (according to (*)) $0 \cdot 999999 \ldots$. Then we must have

$$0 \cdot 999 \ldots 9 \leqslant x < 0 \cdot 999 \ldots 9 + 1/10^n,$$

where there are n 9's each time. Hence

$$1-(1/10^n) \leqslant x < 1,$$

or

$$0 < 1-x \leqslant 1/10^n$$

for all $n \in \mathcal{N}$. By Archimedes condition this is impossible, for since $1-x > 0$ there must exist some n with $1/10^n < 1-x$.

The reason why this sequence of 9's cannot occur is our choice of inequalities in (*). If instead we use

$$a_0 \cdot a_1 a_2 \ldots a_n < x \leqslant a_0 \cdot a_1 a_2 \ldots a_n + 1/10^n \qquad (**)$$

then we get an equally useful definition of the decimal expansion, and it is easy to see that the expansion of the number $x = 1$ now takes the form $0 \cdot 999999 \ldots$.

However, the second rule (**) will now never give us the expansion $1 \cdot 000000 \ldots$.

Comparing (*) and (**), we find that they both give the same decimal expansion for x provided equality never occurs in either of them. Now equality occurs if and only if x is of the form $a_0 \cdot a_1 a_2 \ldots a_n$. (It may look as if we have forgotten $a_0 \cdot a_1 a_2 \ldots a_n + 1/10^n$, but this is equal to $a_0 \cdot a_1 a_2 \ldots (a_n + 1)$ if $a_n \neq 9$, and to whatever we get by carrying if $a_n = 9$, so it amounts to the same thing.) In this case the two methods lead to different results. For instance, if $x = 1\frac{57}{100}$, then the first method gives the decimal expansion $1 \cdot 57000 \ldots$ and the second $1 \cdot 56999 \ldots$. In general, if x has a terminating decimal expansion $x = a_0 \cdot a_1 \ldots a_n$ where $a_n \neq 0$, then the first method gives $a_0 \cdot a_1 a_2 \ldots a_n 000 \ldots$, and the second $a_0 \cdot a_1 a_2 \ldots (a_n - 1)999 \ldots$.

The best way to deal with this situation is to allow *both* possibilities. Simply let $s_n = a_0 \cdot a_1 a_2 \ldots a_n$ be the truncation of the expansion to n decimal places (without rounding off), then let the limit l of the sequence (s_n) be denoted by

$$l = a_0 \cdot a_1 a_2 \ldots a_n \ldots.$$

In this way it can happen that two different sequences have the same limit, precisely when one is a terminating decimal and the other ends in a sequence of 9's as described. For instance if $s_n = 0 \cdot a_1 a_2 \ldots a_n$ where every $a_r = 9$, then the sequence (s_n) has limit denoted by $0 \cdot 99 \ldots 9 \ldots$ This limit is precisely 1, so

$$1 \cdot 00 \ldots 0 \ldots = 0 \cdot 99 \ldots 9 \ldots.$$

It is important not to think that $0 \cdot 99 \ldots 9 \ldots$ is a number 'infinitely smaller' than 1 (whatever that means), they are just two different ways of writing the same real number.

It is convenient to allow both notations because under certain circumstances a calculation may give rise to the infinite sequence of 9's. This will happen using the method given earlier to find the decimal expansion of the limit of a bounded increasing sequence.

EXAMPLE. Suppose $a_1 = 1$ and in general $a_{n+1} = a_n + (\frac{1}{2})^n$, then trivially (a_n) is increasing and a calculation gives $a_n = 2 - (\frac{1}{2})^{n-1}$, so the sequence is bounded above by 2. Using the method of page 32, we find the limit of the sequence (a_n) to be $b_0 \cdot b_1 b_2 \ldots b_n \ldots = 1 \cdot 99 \ldots 9 \ldots$.

Bounded sets

By drawing the picture of a bounded increasing sequence we can actually see the limit process in action, as later terms in the sequence pack together in a rapidly decreasing space. We now consider not just a sequence, but an arbitrary subset $S \subseteq \mathcal{R}$ which is bounded above by some k. This means that $s \leqslant k$ for all $s \in S$. Is there some concept analogous to the limit?

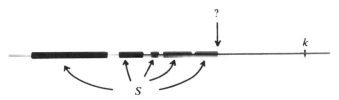

The naive thing to expect is that S has a *greatest element*, a number $s_0 \in S$ such that $s_0 \geqslant s$ for every $s \in S$. Sadly, this is not quite right. For example, if S is the set of all elements of which are strictly less than 1, then there is no element in S which is greater than all the others although S is certainly bounded above (for example by $k = 1$). For suppose that y were a greatest element. Then $y \in S$ so $y < 1$, and then we have

$$y < \tfrac{1}{2}(y + 1) < 1,$$

so $\tfrac{1}{2}(y + 1) \in S$ but is greater than the supposedly greatest element y.

However, all is not lost: we just have to be more subtle. In this example the set S has many upper bounds: in fact any $k \geqslant 1$ is an upper bound for S. Now the set of all upper bounds *does* have a *least* element.

In fact in this example it is 1. In other words, not only is 1 an upper bound, but every other upper bound is bigger.

Before formulating this, let us make absolutely sure that we understand the concepts. A non-empty subset $S \subseteq \mathcal{R}$ is said to be *bounded above* by $k \in \mathcal{R}$ if $s \leq k$ for all $s \in S$. The number k is called an *upper bound* for S.

A subset $S \subseteq \mathcal{R}$ has a *least upper bound* λ if:

(i) λ is an upper bound for S,

(ii) If k is any other upper bound for S, then $\lambda \leq k$.

Although upper bounds are ten a penny, a least upper bound must be unique. For if λ and μ are least upper bounds for S, then (ii), applied to each of them, tells us that $\lambda \leq \mu$ and $\mu \leq \lambda$, and hence that $\lambda = \mu$.

EXAMPLES

(1) If S is the set of all integers, then S has no upper bounds, so certainly no least upper bound.

(2) If S is the set of all real numbers less than or equal to 49, then 49 is the least upper bound of S.

(3) If S is the set of all decimal approximations $1 \cdot 4, 1 \cdot 41, 1 \cdot 414, \ldots$ to $\sqrt{2}$, then the least upper bound of S is $\sqrt{2}$.

(4) If S is the set of all rational numbers r such that $r^2 < 2$, then $\sqrt{2}$ is the least upper bound of S.

In example 2 the least upper bound is an element of S, but in examples 3 and 4 it is not. So even when least upper bounds exist, they may not be members of the original set.

There is once more a parallel set of concepts. A subset S is said to be *bounded below* if there exists $k \in \mathcal{R}$ with $k \leq s$ for all $s \in S$, and k is then called a *lower bound*. The number $\mu \in \mathcal{R}$ is a *greatest lower bound* for S if

(i) μ is a lower bound for S,

(ii) If k is another lower bound for S, then $\mu \geq k$.

A similar trick to that used on decreasing sequences allows us to refer all problems on greatest lower bounds back to least upper bounds. In fact all the basic properties of upper bounds hold for lower bounds, provided we interchange $\geq$ and $\leq$.

We can formulate a more general version of the completeness property of the real numbers:

PROPOSITION 5.   Every non-empty subset of $\mathcal{R}$ which is bounded above has a least upper bound.

'Non-empty' is necessary because any number is an upper bound for a set with no elements. The proof of the above proposition can be made plausible by using decimal expansions in the same sort of way that we dealt with increasing sequences. In fact it is easier to deal with lower bounds and then use the trick to convert to upper bounds. This means we look at:

PROPOSITION 6.   Every non-empty subset of $\mathcal{R}$ which is bounded below has a greatest lower bound.

PROOF.   Let a_0 be the largest integer which is a lower bound. Let a_1 be the largest integer between 0 and 9 for which $a_0 \cdot a_1$ is a lower bound. Then, generally, let a_n be the greatest integer between 0 and 9 for which $a_0 \cdot a_1 a_2 \ldots a_n$ is a lower bound. Then we can show that

$$a_0 \cdot a_1 a_2 \ldots$$

is the greatest lower bound. $\square$

We have remarked that it is not possible to make a drawing sufficiently accurate to distinguish rational numbers from irrational ones. But questions of upper and lower bounds expose a vital theoretical difference between real and rational numbers. Examples (3) and (4) above are bounded sets of rationals with no *rational* least upper bound. In a word: $\mathcal{R}$ is *complete* but $\mathcal{Q}$ is not. It is this property which will play a vital role when we come to a formal definition of the real numbers later in this book.

Exercises

1. Assuming any results you need about prime factorization of natural numbers, show that every positive rational number can be written in exactly one way as a product

$$r = p_1^{\alpha_1} p_2^{\alpha_2} \ldots p_s^{\alpha_s}$$

where $p_1 = 2$, $p_2 = 3, \ldots$ are the primes in increasing order and each α_k is an integer (positive, negative, or zero.)

Write the following rationals in this manner: 14/45, 3/8, 2, 20/45. Show that $\sqrt{r}$ is rational precisely when all of $\alpha_1, \alpha_2, \ldots$ are even. Deduce that for a positive integer n, $\sqrt{n}$ is irrational if and only if n is not the square of an integer.

2. Extend the result of exercise 1 to find those rational numbers r such that $\sqrt[3]{r}$ (cube root of r) is irrational. Show that $\sqrt[n]{(\frac{3}{8})}$ is irrational for all natural numbers $n \geqslant 2$.

3. Which of the following statements are true?
   (a) If x is rational and y is irrational, then $x + y$ is irrational.

(b) If x is rational and y is rational, then $x + y$ is rational.

(c) If x is irrational and y is rational, then $x + y$ is rational.

(d) If x is irrational and y is irrational, then $x + y$ is irrational.

Prove the true ones and give examples to disprove the false ones.

4. Prove that between any two distinct real numbers there exist infinitely many rational numbers and infinitely many distinct irrational numbers. (Here 'infinitely many' means that given any natural number n, there exist at least n numbers with the required property.)

5. For real numbers a, r and natural number n, let $s_n = a + ar + \ldots + ar^n$. Show that $rs_n - s_n = a(r^{n+1} - 1)$ and deduce that

$$\left| s_n - \frac{a}{1-r} \right| = \left| \frac{r^{n+1}}{1-r} \right| \quad \text{for } r \neq 1.$$

For $|r| < 1$ deduce that $s_n \to a/(1-r)$ as $n \to \infty$.

6. Prove that an infinite decimal $x = a_0 \cdot a_1 a_2 a_3 \ldots$ is a rational number if and only if it is 'eventually recurring', that is, after some a_n onwards it repeats the same block of digits indefinitely.

$$x = a_0 \cdot a_1 \ldots a_n \underbrace{a_{n+1} \ldots a_{n+k}} \underbrace{a_{n+1} \ldots a_{n+k}} \underbrace{a_{n+1} \ldots a_{n+k}} \ldots$$

(Hint: One way round, use questions 5 with $a = a_{n+1} \ldots a_{n+k}/10^{n+k}$, $r = 1/10^k$.)

7. Prove that

$$y = 0 \cdot 1234567891011121314151617181920 \ldots.$$

(whose digits are the natural numbers in decimal form, strung end to end) is irrational.

Is

$$0 \cdot 101001000100001 \ldots$$

(where each successive string of 0's has one more digit) rational or irrational?

8. Say whether each of the following sequences (a_n) tends to a limit, and if so, what the limit is. Use the $\varepsilon - N$ definition to prove your answers correct.

    (a) $a_n = n^2$,

    (b) $a_n = 1/(n^2 + 1)$,

    (c) $a_n = 1 + \frac{1}{3} + \frac{1}{9} + \ldots + (\frac{1}{3})^n$,

    (d) $a_n = (-1)^n$,

    (e) $a_n = (-\frac{1}{2})^n$.

PART II

The beginnings of formalization

THE next five chapters develop the techniques we need to place mathematical reasoning on a firmer logical basis. We still permit the use of our intuitive ideas, but now only as *motivation* for the concepts introduced, and no longer as an integral part of the reasoning.

In the 'building' metaphor, we are getting together the bricks, cement, timber, tiles, pipes, and other materials; and assembling a workforce of bricklayers, plasterers, joiners, and plumbers to put them together in the right way. In the 'plant' metaphor it is a question of flowerpots, stakes, forks, and trowels, and a good stock of insecticide to keep the bugs off.

We concentrate on two main ideas: the use of set theory as a source of raw material, and the use of mathematical logic to ensure that the proofs of theorems are rigorous and sound. There are three chapters on sets and related topics, followed by two on logic. We approach both from the point of view of a practical mathematician who is more interested in using them to do mathematics than in their own internal workings.

3

Sets

IN accordance with our point of view, as stated in chapter 1, we shall make no attempt to give a precise *definition* of the concept 'set'. This will not prevent us from explaining what a set is. A set is any collection of objects whatsoever. The word 'collection' is not intended to imply anything about the number of objects in the set: it may be finite or infinite, there may be just one object, or even none. Nor is there any intention to imply any uniformity in the type of object used to make up the set: a perfectly good set might consist of three numbers, two triangles and a function. Obviously such a broad concept allows vast scope for whimsical examples. However, the sets of interest in mathematics will necessarily be those consisting of mathematical objects. At an elementary level one encounters sets of numbers, sets of points in the plane, sets of geometrical curves, sets of equations. In more advanced mathematics, one finds an enormous variety of sets; in fact almost all the concepts of interest are built up from a set-theoretical standpoint.

Nowadays the concept 'set' is considered to be fundamental to the whole of mathematics—even more fundamental than the concept 'number' which earlier ages plumped for. There are many reasons for this. One is that the solution of equations usually yields a set of solutions, rather than just one: quadratic equations, for example, usually have two solutions. Again modern mathematics places emphasis on generality. Interesting theorems tend to apply to a variety of cases. Pythagoras' Theorem is important, not because it applies to one particular right-angled triangle, but because it applies to *all* of them. It thereby expresses a property of the set of all right-angled triangles. The concept of a 'group' (which we will describe later in the book) appears in many guises throughout the whole of mathematics. The language of sets helps us to formulate general properties of a group which will therefore apply to all its appearances. It is this power of expression of general concepts using set-theoretic language that gives modern mathematics its distinctive flavour.

To deal with all the sets which arise in mathematics, it is easiest to develop first the general properties common to all sets, and then to apply them in more special situations. For the rest of this chapter we will concentrate on various natural ways of combining and modifying sets to form other sets. The systematic study of these leads to a kind of 'algebra' of sets, in the same way that a systematic study of the general properties of numbers and the operations of addition, subtraction, multiplication, division, and such leads to an algebra of numbers.

Members

The objects which together make up a given set are called the *members* or *elements* of the set. The members themselves are said to *belong* to the set. To express symbolically that an element x belongs to a set S, we write

$$x \in S.$$

If x does not belong to S, we write

$$x \notin S.$$

In order to know which set is under consideration, it is clear that we must know exactly which objects are members. Conversely, if we know the exact membership, we know which set the members form. Being pedantic about this is not as silly as it might seem, because we often may describe a set in different ways, as then we know we are dealing with one and the same set by looking at its members. For example, if A is the set of solutions of the equation

$$x^2 - 6x + 8 = 0,$$

and B is the set of even integers between 1 and 5, then A and B both have precisely two members, 2 and 4. This means that A and B are the same set. It is sensible, therefore, to say that two sets are *equal* if they have the same members. Equality of two sets S and T will be expressed in the usual way by

$$S = T,$$

and if S and T are not equal, we write

$$S \neq T.$$

This apparently trite criterion for the equality of sets has some interesting consequences, as we shall see in a moment.

The simplest way to specify a set is to list its members (if that is possible). The standard notation is to enclose the list in braces { }. So if we write

$$S = \{1, 2, 3, 4, 5, 6\}$$

we mean that S is the set whose members are the numbers 1, 2, 3, 4, 5, 6, *and only these*. As another example, if

$$T = \{79, \pi^2, \sqrt{(5 + \sqrt{7})}, \tfrac{4}{5}\},$$

then the members of T are the numbers 79, π^2, $\sqrt{(5 + \sqrt{7})}$, and $\tfrac{4}{5}$.

Two points about this notation should be emphasized, both consequences of our notion of equality of sets. First, it is immaterial in what order we write the list of members. The set $\{5, 4, 3, 2, 6, 1\}$ is the *same* as the set S above, and so is the set $\{3, 5, 2, 1, 6, 4\}$. Why? Because in all three cases, we have the same members, namely 1, 2, 3, 4, 5, and 6. The order within the braces arises not from any mathematical cause, but from our conventions about writing from left to right. Second, if elements are repeated in the list, this does not alter the set either. For instance $\{1, 2, 3, 4, 6, 1, 3, 5\}$ is just our old friend S again. Once more, there is a reason for this seemingly peculiar convention. We might combine two lists to give the set consisting of, say, all the proper divisors of 12, namely 1, 2, 3, 4, 6, together with the odd numbers less than 6, which are 1, 3, 5. Just writing one list after the other gives precisely what we have written. In this case it would have been quite easy to go through and cross out repeats, but in general, it is better to retain the flexibility of notation and allow repeats to happen. It follows from our convention about a set being specified by its members, that all the various ways of specifying S have precisely the members 1, 2, 3, 4, 5, 6 and no others.

These remarks may be thought of as pointing out peculiarities of the notation and have no great conceptual significance. We are used to the fact that in writing fractions, say, we have different symbols for the same number: $\tfrac{1}{2} = \tfrac{2}{4} = \tfrac{3}{6}$, etc. In fact this is one of the most common usages of the equality sign, namely if we write $x = y$, we mean that the two symbols on either side of the sign are none other than two different names for the same thing, $2 + 2 = 12 \div 3 = 5 - 1 = +\sqrt{16} = 4$. We use this same convention when we write $S = T$ for equality of sets. Having understood this, there is no essential difficulty here; we have just raised these various questions to dispose of them.

In specifying a set, it may not be convenient, or even possible, to write down a complete list of all the members. The set of prime numbers is better described precisely by that phrase, rather than by the list

$$\{2, 3, 5, 7, 11, 13, 17, 19, \ldots\}.$$

Indicating a few terms of an infinite set in this manner is open to the same sort of misinterpretation as writing the first few terms of a sequence, only slightly worse. A sequence is thought of *in order*, but according to our conventions about sets, the elements inside braces are *not* in any specific order. So the list above might also be written

$$\{7, 17, 37, 47, 2, 11, 3, 5, \ldots\}.$$

Who could sort out this jumble and say with his hand on his heart, that this is the set of all primes? Having made this remark, we must admit that there are occasions when mathematicians use the bracket notation for infinite sets. We do so ourselves sometimes.

In the given case we could be more precise by writing

$$P = \{\text{all prime numbers}\}$$

which is self-explanatory. A slight variation on this, which is very useful, is

$$P = \{p \mid p \text{ is a prime number}\}.$$

Here the braces are read as 'the set of all . . .', the vertical line as 'such that' and then the whole symbol reads 'the set of all p such that p is a prime number', which obviously means 'the set of all prime numbers'. In general a definition of the type

$$Q = \{x \mid \text{something or other involving } x\}$$

means that Q is the set of all x for which the something or other involving x is true. To see how useful this notation is, suppose we want to define S to be the set of solutions of the quadratic equation

$$x^2 - 5x + 6 = 0.$$

We could, of course, solve, and define $S = \{2, 3\}$. Much easier, since it avoids working out the equation, is to write

$$S = \{x \mid x^2 - 5x + 6 = 0\}.$$

This gives a precise and unequivocal definition of S. Of course it is no help in solving the equation! But that is the point of the whole exercise: we can specify the set S without actually doing any calculations.

There is room for ambiguity in this notation which we must face squarely. If we are thinking about integers only then the set

$$\{x \mid 1 \leqslant x \leqslant 5\}$$

consists of the numbers 1, 2, 3, 4, 5, but if we are thinking about real numbers, then all the other real numbers between 1 and 5 are included. The best way out of this is to specify some set Y from which the elements are to be chosen. The notation

$$X = \{x \in Y \mid \text{something or other involving } x\}$$

simply means that X is the set of those members x in the given set Y such that something or other involving x is true. In fact this is just the same as

$$X = \{x \mid x \in Y \text{ and something or other involving } x\},$$

but the first of these notations is preferable since it emphasizes the role of the set Y.

If $\mathscr{Z}$ is the set of integers and $\mathscr{R}$ the set of real numbers, then

$$\{x \in \mathscr{Z} \mid 1 \leqslant x \leqslant 5\}$$

has members 1, 2, 3, 4, 5, whilst every $a \in \mathscr{R}$ satisfying $1 \leqslant a \leqslant 5$ is a member of

$$\{x \in \mathscr{R} \mid 1 \leqslant x \leqslant 5\}.$$

There is an even more serious reason for specifying a set Y from which the members of the set X are chosen, and that is in terms of making sure that the 'something or other involving x' makes sense for all $x \in Y$. The 'something or other' needs to be a property which is clearly true or false for every $x \in Y$, then the set X selected by this property are just those members of Y for which the property is true.

In English grammar, a sentence is divided into two parts, the *subject* of the sentence, and the rest, called the *predicate*, which tells us about the subject.

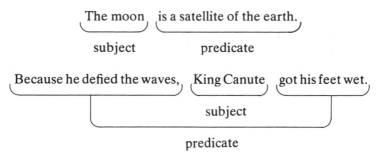

A mathematician, used to using a symbol like x to denote an unknown, might perform the analysis by saying that the predicate in the first sentence is

x is a satellite of the earth

and the predicate in the second is

Because he defied the waves, x got his feet wet.

The beauty of this description is that it specifies the position of the subject in the sentence. To get the original sentence back again we simply substitute the appropriate subject in place of x.

This gives us the idea of a predicate in mathematics. It is simply a sentence involving a symbol x so that when we substitute an element $a \in Y$ for x, then the resultant statement is either clearly true or clearly false. We say that the predicate is 'valid for the set Y' if this is so. For instance the sentence

$$1 \leqslant x \leqslant 5$$

is a predicate which is valid for the set $\mathscr{Z}$. It is also valid for the set $\mathscr{R}$. Substitute any integer or real number and we get a statement which is true or false.

$$1 \leqslant 3 \leqslant 5 \quad \text{is true}$$

$$1 \leqslant 57 \leqslant 5 \text{ is false}$$

and so on.

The set $\{x \in \mathscr{Z} \mid 1 \leqslant x \leqslant 5\}$ is just the set of $x \in \mathscr{Z}$ such that the predicate $1 \leqslant x \leqslant 5$ is true.

A predicate need not be restricted just to sets of numbers. For instance if T is the set of triangles in the plane, then the sentence

x is right-angled

is a predicate valid for the set T and

$$\{x \in T \,|\, x \text{ is right-angled}\}$$

is just the set of right-angled triangles in the plane.

We could go on giving examples galore of predicates, but there will be plenty turning up in the text anyway. The reader should make it clear in his own mind that whenever the symbolism

$$\{x \in Y \,|\, P(x)\}$$

is used, then $P(x)$ is a predicate in x which is valid for all $x \in Y$.

Subsets

Within any given set A there exist other sets, obtained by omitting some of the elements. These are called subsets of A. More formally, we say that B is a *subset* of A if every element of B is an element of A, and write

$$B \subseteq A$$

or

$$A \supseteq B.$$

We also say that B is *contained in*, or *included in*, A. With this definition we have $A \subseteq A$, for trivial reasons. If $B \subseteq A$ and $B \neq A$ then we say that B is a *proper subset* of A and write†

$$B \subsetneqq A.$$

The criterion for equality of sets leads to a trivial but useful result:

PROPOSITION 1.   Let A and B be sets. Then $A = B$ if and only if $A \subseteq B$ and $B \subseteq A$.

PROOF.   If $A = B$ then, since $A \subseteq A$, it follows that $A \subseteq B$ and $B \subseteq A$. Conversely, suppose $A \subseteq B$ and $B \subseteq A$. Then each element of A is an element of B, and each element of B is an element of A. Hence A and B have the same elements, so $A = B$.   □

The practical use of this proposition is that we often wish to show the equality of two sets where each may be given in terms of some predicate or other. To simplify matters we start with a typical element in A (given in terms of the appropriate predicate) and show that this

† Many mathematicians use $\subset$ where we have used $\subseteq$ and others write $\subset$ where we have chosen $\subsetneqq$. Fortunately the use of $\subseteq$ is totally unambiguous.

element is also a member of B. This verifies $A \subseteq B$, then we do a suitable argument of the same nature to show that $B \subseteq A$. We will see plenty of examples of this type of procedure very soon (in propositions 3, 4, and 5, for instance).

A basic property of subsets is that a subset of a subset is itself a subset:

PROPOSITION 2.   If A, B, C are sets with $A \subseteq B$ and $B \subseteq C$, then $A \subseteq C$.

PROOF.   Every element of A is an element of B and every element of B is an element of C. Therefore every element of A is an element of C, so $A \subseteq C$.   □

It is highly important not to confuse subsets and members, as the two concepts are quite different. The *members* of $\{1, 2\}$ are 1 and 2. The subsets of $\{1, 2\}$ are $\{1, 2\}$, $\{1\}$, $\{2\}$, and a fourth subset which for the moment is best written as $\{\ \}$.

Further, proposition 2 becomes false if we change '$\subseteq$' to '$\in$'. Members of members need not be members. For example, let $A = 1$, $B = \{1, 2\}$, $C = \{\{1, 2\}, \{3, 4\}\}$. Then $A \in B$ and $B \in C$. But the members of C are $\{1, 2\}$ and $\{3, 4\}$, so $A = 1$ is not a member.

Now let us return to that set $\{\ \}$. We say that a set is *empty* if it has no members. For instance the set

$$\{x \in \mathscr{Q} \mid x = x + 1\}$$

is empty, because the equation $x = x + 1$ has no solutions in $\mathscr{Q}$. An empty set has remarkable properties (remarkable at first sight, that is) by default. For instance if E is an empty set and X is any set whatsoever, then $E \subseteq X$. Why? We have to show that every element of E is an element of X. The only way that this can fail is if E has some element e which does not belong to X. But E, being empty, has no elements at all, so cannot contain any such element.

This (curious but logical) argument is called 'vacuous reasoning' and should not be thought of as being particularly significant.

Suppose we have two empty sets E and E'. The above tells us that $E \subseteq E'$ and $E' \subseteq E$. Then proposition 1 tells us that $E = E'$. *All empty sets are equal.* Hence there is a *unique* empty set. We therefore give it a special symbol: we write

$$\varnothing$$

to denote *the* empty set.

This is hardly surprising. In the absence of any elements whatsoever, we have no way to distinguish two empty sets. In the words of [15], 'the contents of two empty paper bags are equal'.

Is there a universe?

Just as there is an empty set $\varnothing$ which contains no elements, we might ask whether there is a very large set Ω which includes absolutely everything. This turns out to be far too fanciful. Such a set would have to be an incredibly vast rag-bag; if it contained everything it would include all numbers, all elements of every set, all sets, all places in the universe, the Declaration of Independence, Winston Churchill, the year 1066, the wit of Oscar Wilde, ... If we ever dare contemplate such an Ω, then Ω itself must be an acceptable concept and we would have to include it in the collection of absolutely everything. So we find a set satisfying $\Omega \in \Omega$! Now most sensible sets do not belong to themselves, in fact the reader could while away an interesting half-hour trying to find such a set. If we select from the putative set Ω everything which is a set which doesn't belong to itself, we get:

$$S = \{A \in \Omega \mid A \notin A\}.$$

Now ask the $64,000 question. Does $S \in S$? ...
If $S \in S$, then, according to the defining predicate, $S \notin S$.
If $S \notin S$, then S satisfies the defining predicate, so $S \in S$!
Our flight of fancy in assuming the existence of a universe Ω has led us into a paradox. There cannot be a universal set.

It might be thought that we can salvage the situation by removing all the whimsical things and concentrating our attention on a universal set in the realms of mathematics. This too has its pitfalls. If we try to contemplate a set Ω_M of all mathematical objects (whatever that means), then we would arrive in the same cleft stick when we consider the subset of Ω_M consisting of all mathematical objects which don't belong to themselves.

To avoid such paradoxes, it is essential that the sets we consider are clearly defined and we know precisely which objects are members and which are not.

The non-existence of a universal set is another reason why the notation

$$\{x \in Y \mid P(x)\},$$

where Y is a known set and $P(x)$ is a predicate, is preferable to

$$\{x \mid P(x)\}.$$

Having specified a set Y we can investigate the predicate $P(x)$ and make sure it is valid for all elements of Y before selecting those elements in Y for which the predicate is true. Used indiscriminately the notation $\{x \mid P(x)\}$, which allows us to try absolutely any object x to see if it is a member, is like considering $\{x \in \Omega \mid P(x)\}$. We have seen that there is no universal set Ω. If we don't specify a set Y in the first place, then we have unlimited choice of objects to try in the predicate $P(x)$. We might then find someone considering an element which hadn't been intended at the outset and end up in a paradoxical situation again. Let us consider a particular example to make this clear. If $\mathscr{Z}$ is the set of integers, $\mathscr{R}$ the set of real numbers, and T the set of all triangles in the plane, then $\mathscr{Z} \notin \mathscr{Z}$, $\mathscr{R} \notin \mathscr{R}$, $T \notin T$. If Y is the set whose members are $\mathscr{Z}$, $\mathscr{R}$, T, then

$$\{x \in Y \mid x \notin x\} = \{\mathscr{Z}, \mathscr{R}, T\}.$$

On this set Y, the property $x \notin x$ is a perfectly acceptable predicate. If we consider

$$\{x \mid x \notin x\},$$

then, with no restrictions on x, the imagination can run riot, and considering $S = \{x \mid x \notin x\}$ itself, we end up with the same contradiction that we found before, $S \in S$ if and only if $S \notin S$.

The moral of this particular discussion is that set theory is a system of notation, not a magic prescription. As such it is as good as the manner in which it is used. With good sense, it behaves well. But like any other system, use it badly then things can go wrong.

Union and intersection

Two important methods for 'combining' sets are known as the union and intersection. The *union* of the sets A and B is the set whose elements are those of A together with those of B. If

$$A = \{1, 2, 3\}$$

$$B = \{3, 4, 5\}$$

then the union is $\{1, 2, 3, 4, 5\}$. We write $A \cup B$ to denote the union of A and B. We then have

$$A \cup B = \{x \mid x \in A \text{ or } x \in B \text{ (or both)}\}.$$

The intersection of A and B is the set whose elements belong both to A and to B. With A and B above, their intersection is $\{3\}$, because only 3 belongs to both of them. The symbol for the intersection is $A \cap B$. In this case,

$$A \cap B = \{x \mid x \in A \text{ and } x \in B\}.$$

Notice that the intersection can also be written as

$$A \cap B = \{x \in A \mid x \in B\},$$

so that we can think of it being the subset of A selected using the predicate $x \in B$. (Equivalently we could think of it as being the subset of B which satisfies the predicate $x \in A$.) The union, on the other hand, involves constructing a new set which is (usually) bigger than both A and B, so here we have an example of set construction which doesn't select elements from a previously prescribed set Y.

The operations of union and intersection obey certain 'standard' laws. Most of these are obvious, but for convenience we list them in the next three propositions.

PROPOSITION 3.   Let A, B, C be sets. Then
(a) $A \cup \emptyset = A$
(b) $A \cup A = A$
(c) $A \cup B = B \cup A$
(d) $(A \cup B) \cup C = A \cup (B \cup C)$.

PROOF. Only (d) is remotely difficult. Suppose that $x \in (A \cup B) \cup C$. Then either $x \in A \cup B$ or $x \in C$. If $x \in C$, then $x \in B \cup C$, so $x \in A \cup (B \cup C)$. If not, then $x \in A \cup B$, so either $x \in A$ or $x \in B$. In each case we again have $x \in A \cup (B \cup C)$. So we have proved that if $x \in (A \cup B) \cup C$ then $x \in A \cup (B \cup C)$, or that

$$(A \cup B) \cup C \subseteq A \cup (B \cup C).$$

A similar, equally involved and equally trivial, argument shows that

$$A \cup (B \cup C) \subseteq (A \cup B) \cup C.$$

Using proposition 1 we obtain equality as desired.   □

It should not have escaped the reader's attention that this proof is more complicated that the situation really warrants, because it is obvious that, say, $(A \cup B) \cup C$ is the set whose members are those of A, those of B, and those of C together; and this is the same set as $A \cup (B \cup C)$. Once we know this, it is possible to omit the brackets altogether and write just

$$A \cup B \cup C.$$

Similar results hold for intersections:

PROPOSITION 4.
(a) $A \cap \varnothing = \varnothing$
(b) $A \cap A = A$
(c) $A \cap B = B \cap A$
(d) $(A \cap B) \cap C = A \cap (B \cap C)$.

The proofs are analogous to those in theorem 3.   □

Finally, there are two equations which mix up unions and intersections:

PROPOSITION 5.
(a) $A \cup (B \cap C) = (A \cup B) \cap (A \cup C)$
(b) $A \cap (B \cup C) = (A \cap B) \cup (A \cap C)$.

PROOF.   Let $x \in A \cup (B \cap C)$. Then either $x \in A$ or $x \in B \cap C$. If $x \in A$ then certainly $x \in A \cup B$ and $x \in A \cup C$, hence $x \in (A \cup B) \cap (A \cup C)$. Alternatively, $x \in B \cap C$ gives $x \in B$ and $x \in C$. Hence $x \in A \cup B$ and $x \in A \cup C$, so $x \in (A \cup B) \cap (A \cup C)$. This proves that

$$A \cup (B \cap C) \subseteq (A \cup B) \cap (A \cup C). \qquad (*)$$

Conversely, suppose $y \in (A \cup B) \cap (A \cup C)$. Then $y \in A \cup B$ and $y \in A \cup C$. There are two cases to consider: when $y \in A$ and when $y \notin A$. If $y \in A$, then certainly $y \in A \cup (B \cap C)$. On the other hand, if $y \notin A$ then, since $y \in A \cup B$, we must have $y \in B$; similarly $y \in C$. Thus $y \in B \cap C$, which again implies $y \in A \cup (B \cap C)$. Therefore

$$(A \cup B) \cap (A \cup C) \subseteq A \cup (B \cap C).$$

Together with (*), this yields the desired result.
The proof of (b) is analogous.   □

Theorem 5 is a pair of 'distributive laws' which should be compared with the way that multiplication of numbers is distributive over addition:

$$a \, . \, (b + c) = (a \, . \, b) + (a \, . \, c).$$

With numbers, however, the interchange of the two operations does not give a new rule:

$$a + (b \, . \, c) = (a + b) \, . \, (a + c)$$

is *not* true in general.

The operations ∪ and ∩ on sets behave in a much more symmetrical way: each is distributive over the other.

One way to visualize these various set theoretic identities is to draw so-called *Venn diagrams*. The identity

$$A \cup (B \cap C) = (A \cup B) \cap (A \cup C)$$

can be represented by drawing three overlapping discs, supposed to represent the sets A, B, C, and proceeding as follows:

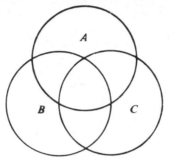

$B \cap C$ is the shaded region common to B and C:

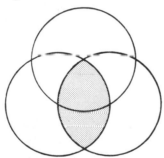

and the union of this with A is:

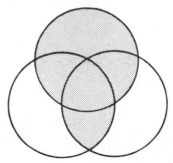

On the other hand, $A \cup B$ is:

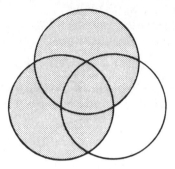

and $A \cup C$ is:

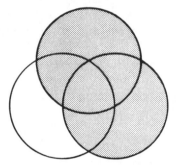

so $(A \cup B) \cap (A \cup C)$ is the region common to both, which is:

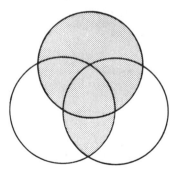

This is the same as before.

The reader may wish to try his hand at illustrating the other identities in theorems 3, 4, and 5 by drawing the appropriate Venn diagrams. Such pictorial devices, if well chosen, aid most people to get a coherent idea of what is going on. To get the most general picture, the diagram must be drawn with care. With one set A, there are two distinct regions involved, inside A and outside:

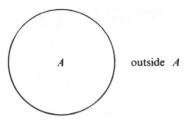

With two sets, A and B, there are four regions, outside both, inside A but not B, inside B but not A, and inside both of them:

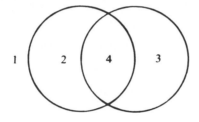

With three sets, A, B, and C, there are eight regions:

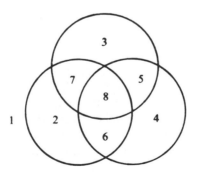

If we add a fourth set D so that D meets each of these eight regions and the area outside D meets each of the regions, then we get sixteen regions in all. There is no way that this can be achieved if A, B, C, and D are all drawn as circles. Just try to draw a fourth circle in the last diagram above to meet this prescription and you will see what we mean. It can be done, but not with a circle:

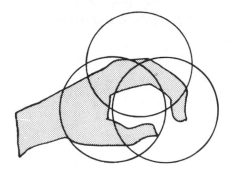

Venn himself was aware of such limitations when he first drew the diagrams. They can be ironed out by using odd shapes to represent the sets, but in view of these technicalities, it is better to consider the pictures as an aid to one's mental processes, rather than an actual method of proof. A proof about a general relationship between sets is best undertaken in a manner analogous to those in propositions 3, 4, and 5.

For instance, there is a general connection between unions, intersections, and subsets:

PROPOSITION 6. If A and B are sets, the following are equivalent:
  (a) $A \subseteq B$
  (b) $A \cap B = A$
  (c) $A \cup B = B$.

PROOF.   Equation (b) says that the elements common to A and B are all the elements of A, so every element of A belongs to B, which implies $A \subseteq B$. The converse is obvious, so (a) and (b) are equivalent.

Equation (c) says that if we add to B the elements of A, we still get B. Therefore no element of A can fail to belong to B, and again $A \subseteq B$. The converse is once more obvious, so (a) and (c) are equivalent. □

Complements

Let A and B be sets. The set-theoretic *difference* $A \setminus B$ is defined to be the set of all those elements of A which do not belong to B. In symbols,

$$A \setminus B = \{x \in A \mid x \notin B\}.$$

In a Venn diagram, $A \backslash B$ is the shaded region in:

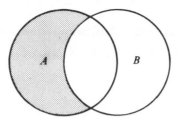

If B is a subset of A, then we call $A \backslash B$ the *complement of B relative to A*.

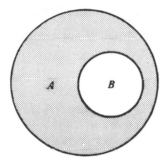

It would be nice to forget about A entirely, thus defining the complement of B to consist of everything not belonging to B. This is too much to ask because it would mean that B and its supposed complement would make up a set Ω which contains absolutely everything and we have already shown that such a set cannot exist. In a particular piece of mathematics there may be a set U around which includes all the elements we wish to consider. We will call this set the *universe of discourse* or *universal set* (universal that is for its current purpose). When dealing with the set of integers, for example, we might take the universal set to be $U = \mathscr{Z}$. Of course, $U = \mathscr{R}$ would do just as well. The important thing is that the universal set should be sufficiently all embracing to include all the elements under discussion. As one of us says elsewhere 'in a discussion about dogs, when thinking about all non-sheepdogs, it is pointless to worry about camels'.

Having agreed upon U, we define the *complement* B^c of every subset B of U by

$$B^c = U \backslash B.$$

Thus B^c is the complement of B relative to U. But because U is agreed upon, we can omit it from the notation, which is the object of the exercise.

Of course, the operation ^{c} obeys some simple laws, including:

PROPOSITION 7.   If A and B are subsets of the universal set U, then
(a) $\varnothing^c = U$
(b) $U^c = \varnothing$
(c) $(A^c)^c = A$
(d) If $A \subseteq B$ then $A^c \supseteq B^c$.   □

In view of (c) we can write $A^{cc} = (A^c)^c = A$.

Less elementary, but highly interesting are *De Morgan's Laws*:

PROPOSITION 8.   If A and B are subsets of the universal set U, then:
(a) $(A \cup B)^c = A^c \cap B^c$
(b) $(A \cap B)^c = A^c \cup B^c$.

PROOF.   Let $x \in (A \cup B)^c$. Then $x \notin A \cup B$. This implies that $x \notin A$ and $x \notin B$, so $x \in A^c$ and $x \in B^c$, so $x \in A^c \cap B^c$. Therefore $(A \cup B)^c \subseteq A^c \cap B^c$. To obtain the reverse inclusion, reverse the steps in the argument.

This proves (a). Equation (b) can be proved similarly. Instead, we could replace A by A^c and B by B^c in (a), which gives

$$(A^c \cup B^c)^c = A^{cc} \cap B^{cc} = A \cap B.$$

Taking complements, we have

$$A^c \cup B^c = (A^c \cup B^c)^{cc} = (A \cap B)^c$$

and this is (b).   □

These laws explain a phenomenon which the alert reader may have observed: the set-theoretic laws come in pairs, so that if we start with one and change all unions to intersections and all intersections to unions we obtain another. We could formulate this as the:

De Morgan duality principle

If in any valid set-theoretic identity involving only the operations $\cup$ and $\cap$, the operations $\cup$ and $\cap$ are interchanged throughout, the result is another valid identity.

To prove this in general is not hard, but it needs a complicated induction argument which completely hides the basic idea. The following is a typical case. Start with the identity

$$A \cup (B \cap C) = (A \cup B) \cap (A \cup C).$$

Take complements of both sides and use De Morgan's laws to get

$$A^c \cap (B \cap C)^c = (A \cup B)^c \cup (A \cup C)^c,$$

then use De Morgan again to get

$$A^c \cap (B^c \cup C^c) = (A^c \cap B^c) \cup (A^c \cap C^c).$$

Already we have interchanged $\cup$ and $\cap$. Now systematically replace A by A^c, B by B^c, C by C^c. Since the equation is true for *any* sets A, B, C this is legitimate. We get

$$A \cap (B \cup C) = (A \cap B) \cup (A \cap C).$$

This is the original law, with $\cup$'s and $\cap$'s interchanged.

QUESTION.   How does the presence of the operation ^{c} affect the argument? (Try the identity

$$B \cup (A \cap A^c) = B$$

and use the same approach. What happens?)

Sets of sets

It may happen that all the elements of a given set S are themselves sets. Indeed, it is often a useful device to consider a set of sets. For instance we may have $S = \{A, B\}$ where $A = \{1, 2\}$, $B = \{2, 3, 4\}$. A more sophisticated example is to take any set X and let $\mathbb{P}(X)$ be the set of all subsets of X. This is called the *power set of X* and satisfies the property:

$$Y \in \mathbb{P}(X) \text{ if and only if } Y \subseteq X.$$

For example, if $X = \{0, 1\}$, then $\mathbb{P}(X) = \{\emptyset, \{0\}, \{1\}, \{0, 1\}\}$. In cases like these, where every member of S is itself a set, we can go a level further and consider the elements belonging to these members. This gives us generalizations of the notions of union and intersection:

$$\bigcup S = \{x \mid x \in A \text{ for some } A \in S\},$$

$$\bigcap S = \{x \mid x \in A \text{ for every } A \in S\}.$$

We call $\bigcup S$ the 'union of S' and $\bigcap S$ the 'intersection of S'. Put into words, the union of S consists of all the elements in the members of S

and the intersection of S consists of those elements common to all members of S. For instance

$$\bigcup\{\{1, 2\}, \{2, 3, 4\}\} = \{1, 2, 3, 4\}$$
$$\bigcap\{\{1, 2\}, \{2, 3, 4\}\} = \{2\}.$$

In general, for any set X, we have

$$\bigcup\mathbb{P}(X) = X$$
$$\bigcap\mathbb{P}(X) = \varnothing.$$

Although this notation may seem a little strange at first, it is extremely economical and it does act as a genuine extension of the usual concepts. For instance, given two sets A_1, A_2, let $S = \{A_1, A_2\}$, then

$$\bigcup S = A_1 \cup A_2$$
$$\bigcap S = A_1 \cap A_2.$$

More generally

$$\bigcup\{A_1, A_2, \ldots, A_n\} = A_1 \cup A_2 \cup \ldots \cup A_n$$
$$\bigcap\{A_1, A_2, \ldots, A_n\} = A_1 \cap A_2 \cap \ldots \cap A_n.$$

An alternative (and much more used) notation for these last two concepts is to write

$$A_1 \cup A_2 \cup \ldots \cup A_n = \bigcup_{r=1}^{n} A_r$$

$$A_1 \cap A_2 \cap \ldots \cap A_n = \bigcap_{r=1}^{n} A_r.$$

We shall return to generalized unions and intersections once more at the end of chapter 5.

Exercises

1. Which of the following sets are equal?
   (a) $\{-1, 1, 2\}$,
   (b) $\{-1, 2, 1, 2\}$,
   (c) $\{n \in \mathscr{Z} \mid |n| \leqslant 2 \text{ and } n \neq 0\}$,
   (d) $\{2, 1, 2, -2, -1, 2\}$,
   (e) $\{2, -2\} \cup \{1, -1\}$,
   (f) $\{-2, -1, 1, 2\} \cap \{-1, 0, 1, 2, 3\}$.

2. Prove that for all sets A, B,

$$(A \backslash B) \cup (B \backslash A) = (A \cup B) \backslash (A \cap B).$$

If A is the set of even integers, and B is the set of integers which are multiples of 3, describe the set $(A \backslash B) \cup (B \backslash A)$.

3. Write out the proofs of propositions 3(a), 3(b), 3(c), and all of proposition 4. Draw Venn diagrams to illustrate these results.

4. Draw a Venn diagram suitable for all formulae involving five different sets.

5. If $S = \{$all subsets $X \subseteq \mathscr{Z}$ such that $0 \in X\}$, find $\bigcap S$, $\bigcup S$.

6. If $S = S_1 \cup S_2$, prove that $\bigcup S = (\bigcup S_1) \cup (\bigcup S_2)$.

7. If A has n elements ($n \in \mathscr{N}$), calculate the number of subsets of A. If you are acquainted with proof by induction, prove your result by this technique.

8. If A, B, C are finite sets and $|A|$ denotes the number of elements in A, show that

$$|A \cup B \cup C| = |A| + |B| + |C| - |A \cap B| - |B \cap C| - |C \cap A| + |A \cap B \cap C|.$$

Draw a Venn diagram.

9. In each of the following statements, if we replace S by one of $\mathscr{N}$, $\mathscr{Z}$, $\mathscr{Q}$, $\mathscr{R}$, then we get a true statement. Find the appropriate set in each case:
(a) $\{x \in S | x^3 = 5\} \neq \varnothing$,
(b) $\{x \in S | -1 \leqslant x \leqslant 1\} = \{1\}$,
(c) $\{x \in S | 2 < x^2 < 5\} \backslash \{x \in S | x > 0\} = \{-2\}$,
(d) $\{x \in S | 1 < x \leqslant 4\} = \{x \in S | x^2 = 4\} \cup \{3, 4\}$,
(e) $\{x \in S | 4x^7 - 1\} \backslash \{x \in S | x < 0\} = \{x \in S | 5x^2 = 3\} \cup \{x \subset S | 2x - 1\} \neq \varnothing.$

10. The equation $x + y = z$ has many solutions $x, y, z \in \mathscr{N}$; the equation $x^2 + y^2 = z^2$ has solutions including $x = 3$, $y = 4$, $z = 5$.

Let $F = \{n \in \mathscr{N} | x^n + y^n = z^n$ has a solution where $x, y, z \in \mathscr{N}\}$.

What must be done to show $F = \{1, 2\}$? What does this tell us about verifying equality between sets in general?

4

Relations

THE aim of this chapter is to introduce one of the most important concepts in set theory. The notion of a relation is one that is found throughout mathematics and applies in many situations outside the subject as well. Examples involving numbers include 'greater than', 'less than', 'divides', 'is not equal to', examples from the realms of set theory include 'is a subset of', belongs to'; examples from other areas include 'is the brother of', 'is the son of'. What all these have in common is that they refer to two things and the first is either related to the second in the manner described, or not. Thus the statement $a > b$, where a and b are integers, is either true or not true ($2 > 1$ is true, $1 > 2$ is false).

The two things which are related must be taken in a specific order, for instance the statement $a > b$ is quite different from $b > a$, so the first thing that we do in this chapter is to set up some machinery to do with ordered pairs. It is also to be remarked that relations can occur between elements in different sets, in some cases we can have a relation between elements in a set A and those in a set B. Most of the examples we mentioned concern objects from the same set, but we slipped one into the set-theoretic list which was 'belongs to'. If A is a set of elements and B is a set whose members are themselves sets, then we can see whether $x \in Y$ for each member $x \in A$ and each member $Y \in B$. Since $x \in Y$ is either true, or not true, for every $x \in A$ and $Y \in B$, this will be a relation between A and B in the sense to be described in this chapter. The beauty of the description given is that it can be formulated entirely in set-theoretic terms.

At the end of the chapter we develop a detailed theory of two particularly important types of relation: equivalence relations and order relations.

Ordered pairs

We have said that for sets the order in which we write the elements in a list makes no difference, so that for a set with two elements,

$\{a, b\} = \{b, a\}$. This is all very well, but there are occasions on which it is essential to distinguish the order. For instance, in coordinate geometry we think of all the points of the plane as being represented by pairs (x, y) of real numbers. The order is crucial, for example the points $(1, 2)$ and $(2, 1)$ are different:

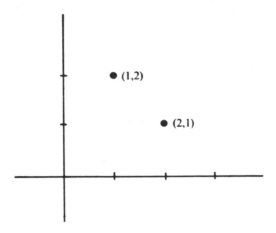

We are thus led to the concept of an *ordered pair* (x, y), round brackets being used to make the distinction from $\{x, y\}$. The important property that we require of this new concept is that

$$(x, y) = (u, v) \quad \text{if and only if} \quad x = u \quad \text{and} \quad y = v. \quad (*)$$

This notion can be used throughout set theory. Given sets A, B then the cartesian product $A \times B$ is the set of all such ordered pairs:

$$A \times B = \{(x, y) | x \in A, y \in B\}.$$

This is all very fine; the only problem is that we haven't actually said precisely what we mean by an ordered pair. What is (x, y)? If $A = B = \mathscr{R}$, then we can think of an ordered pair (x, y) as being a point in the coordinate plane using cartesian coordinates. This is indeed where the notion of ordered pair arose. In this sense we can refer to the plane as $\mathscr{R} \times \mathscr{R}$ (or, in more usual mathematical shorthand, as $\mathscr{R}^2$). But what happens if A is a set like {apple, orange, grapefruit} and B is {knife, fork}, what then is $A \times B$? It certainly consists of the ordered pairs:

   (apple, knife), (apple, fork), (orange, knife), (orange, fork),
   (grapefruit, knife), (grapefruit, fork).

So we return to the question: what is the ordered pair (apple, knife)? The solution lies not in 'what is it?', but in 'how do we get it?'. The answer is that to obtain (x, y) in general, we first select x from A, then we select y from B. The mathematician Kuratowski saw in this process a possible abstract definition of (x, y) using only set-theoretic notions that we have already described. Having selected x from A, we have the singleton set $\{x\}$, then selecting y from B we arrive at the set $\{x, y\}$. Kuratowski simply defined the ordered pair (x, y) as consisting of these sets:

DEFINITION (Kuratowski)   The ordered pair (x, y) of two elements x, y is defined to be the set

$$(x, y) = \{\{x\}, \{x, y\}\}.$$

This peculiar looking definition has the advantage that it satisfies the property (*) which is vital for ordered pairs:

PROPOSITION 1.   With Kuratowski's definition, we have

$$(x, y) = (u, v) \quad \text{if and only if} \quad x = u \quad \text{and} \quad y = v.$$

PROOF   Certainly if $x = u$, $y = v$, then the definition gives $(x, y) = (u, v)$. In the other direction, suppose that $(x, y) = (u, v)$. If $x \neq y$, then $(x, y) = \{\{x\}, \{x, y\}\}$ has two distinct members, $\{x\}$ and $\{x, y\}$ which must each belong to $(u, v) = \{\{u,\}, \{u, v\}\}$. This means that the members $\{u\}$ and $\{u, v\}$ must be different also, implying $u \neq v$. Now we must have $\{x\} = \{u\}$, or $\{x\} = \{u, v\}$ and the latter is clearly impossible (because it would mean that u, v both belonged to $\{x\}$, implying $u = x = v$, contradicting $u \neq v$). So $\{x\} = \{u\}$ and $x = u$. In a similar fashion, $\{x, y\} = \{u, v\}$, and since $x = u$, $x \neq y$ and $y \in \{u, v\}$, we deduce that $y = v$. Thus $x = u$ and $y = v$, as required.

If $x = y$, the set-theoretic construction collapses somewhat to give

$$(x, y) = \{\{x\}, \{x, y\}\} = \{\{x\}, \{x, x\}\} = \{\{x\}, \{x\}\} = \{\{x\}\},$$

so (x, y) has only *one* member, namely $\{x\}$. If $(x, y) = (u, v)$, then (u, v) has only one member also, implying $\{u\} = \{u, v\}$, so $u = v$ and $(u, v) = \{\{u\}\}$. The equality $(x, y) = (u, v)$ then becomes $\{\{x\}\} = \{\{u\}\}$, which reduces successively to $\{x\} = \{u\}$ and then $x = u$. Thus this case reduces to $x = y = u = v$ and the proof is complete.† □

---

† By being a little more sophisticated, we can prove this result much more quickly. In the notation of the last section of chapter 3, we find

$$\bigcap\{\{x\}, \{x, y\}\} = \{x\} \quad \text{and} \quad \bigcup\{\{x\}, \{x, y\}\} = \{x, y\},$$

So $\bigcap(x, y) = \{x\}, \bigcup(x, y) = \{x, y\}$. If $(x, y) = (u, v)$, by comparing intersections and unions we find

$$\{x\} = \{u\}, \qquad \{x, y\} = \{u, v\}.$$

The first gives $x = u$ and from this (whether $x = y$ or not), the second gives $y = v$.

Where does this get us? First the good news: we have a definition of the ordered pair (x, y) involving only established set theoretic concepts. Then the bad news: the definition does not correspond to the intuitive notion of ordered pairs in coordinate geometry. Indeed if any mathematician were asked to visualize (2,1), he would, like as not, think of it as a point in the plane; it is most unlikely that his thoughts would revolve around the idea $\{\{2\}, \{2,1\}\}$.

The pragmatic solution to adopt is to let Kuratowski's definition fade into the background, safe in the knowledge that it is there should one ever be asked to give a rigorous foundation. The important notion to depend on is the property (*):

$$(x, y) = (u, v) \quad \text{if and only if} \quad x = u \quad \text{and} \quad y = v.$$

Given this, we consider $A \times B$ to be the set of all such ordered pairs

$$A \times B = \{(x, y) | x \in A, y \in B\}.$$

The interpretation of $\mathscr{R} \times \mathscr{R}$ as points in the plane remains a most useful one, it certainly satisfies (*) which is its most useful characteristic. This interpretation of $A \times B$ is also useful when A and B are subsets of $\mathscr{R}$. For instance if $A = \{1, 2, 3\}$, and $B = \{5, 7\}$, then $A \times B$ is the set

$$\{(1, 5), (1, 7), (2, 5), (2, 7), (3, 5), (3, 7)\}.$$

Thinking in terms of Cartesian coordinates, we can draw a picture:

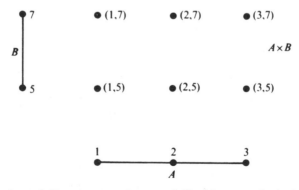

When A and B are not subsets of $\mathscr{R}$ this sort of picture is less appropriate, but it can still be useful. For example if $A = \{a, b, c\}$ and $B = \{u, v\}$, then

$$A \times B = \{(a, u), (a, v), (b, u), (b, v), (c, u), (c, v)\}$$

and the structure is represented by

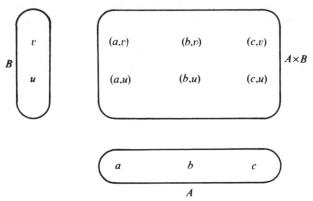

Note that in general, $A \times B \neq B \times A$. For example, with A and B as above, we have

$$B \times A = \{(u, a), (u, b), (u, c), (v, a), (v, b), (v, c)\}$$

which is not the same as $A \times B$. However, the cartesian product does obey some general laws:

PROPOSITION 2.   For any sets A, B, C we have

(a) $(A \cup B) \times C = (A \times C) \cup (B \times C)$

(b) $(A \cap B) \times C = (A \times C) \cap (B \times C)$

(c) $A \times (B \cup C) = (A \times B) \cup (A \times C)$

(d) $A \times (B \cap C) = (A \times B) \cap (A \times C)$.

PROOF.   All are easy, and the argument is similar in each case, so we will prove only (a), leaving the remainder as an exercise.

Let $(u, v) \in (A \cup B) \times C$. Then $u \in A \cup B$, $v \in C$. So $u \in A$ or $u \in B$. If $u \in A$ then $(u, v) \in A \times C$; if $u \in B$ then $(u, v) \in B \times C$. Either way, $(u, v) \in (A \times C) \cup (B \times C)$. Therefore

$$(A \cup B) \times C \subseteq (A \times C) \cup (B \times C).$$

Now let $x = (y, z) \in (A \times C) \cup (B \times C)$. Either $x \in A \times C$ or $x \in B \times C$. In the first case, $y \in A$ and $z \in C$. In the second, $y \in B$ and $z \in C$, so $x = (y, z) \in (A \cup B) \times C$. This shows

$$(A \times C) \cup (B \times C) \subseteq (A \cup B) \times C.$$

Putting the two parts together finishes the proof. $\square$

This can be illustrated by the following diagram.

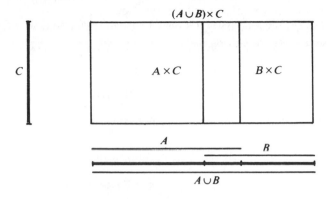

$(A \cup B) \times C$

$A \times C$ $B \times C$

C

A

B

$A \cup B$

PROPOSITION 3.   For all sets A, B, C, D, we have

$$(A \times B) \cap (C \times D) = (A \cap C) \times (B \cap D)$$

PROOF.   Let $x = (y, z) \in (A \times B) \cap (C \times D)$.   Then $y \in A, z \in B$, and $y \in C, z \in D$. So $y \in A \cap C, z \in B \cap D$, so $x \in (A \cap C) \times (B \cap D)$. Hence

$$(A \times B) \cap (C \times D) \subseteq (A \cap C) \times (B \cap D)$$

Conversely let $x = (y, z) \in (A \cap C) \times (B \cap D)$. Then $y \in A$ and $y \in C$, $z \in B$, and $z \in D$, so $x \in (A \times B) \cap (C \times D)$. Therefore

$$(A \cap C) \times (B \cap D) \subseteq (A \times B) \cap (C \times D).$$

Pictorially:

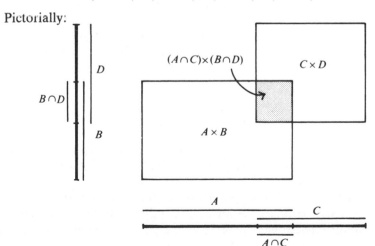

D

$(A \cap C) \times (B \cap D)$ $C \times D$

$B \cap D$

B

$A \times B$

A

C

$A \cap C$

The same picture should make it clear why a theorem like this does *not* hold for unions in place of intersections.

Having got ordered pairs it is easy to go on to define ordered triples, quadruples, etc. by setting

$$(a, b, c) = ((a, b), c)$$

$$(a, b, c, d) = (((a, b), c), d)$$

and so on. These are elements of repeated cartesian products, defined by

$$A \times B \times C = (A \times B) \times C$$

$$A \times B \times C \times D = ((A \times B) \times C) \times D.$$

Later we shall find a better way to obtain the general concept of an ordered n-tuple

$$(a_1, a_2, \ldots, a_n)$$

for any natural number n. At our present stage we can do this for any *particular n* by repeating the process used for triples or quadruples. These generalizations have similar properties to the main property (*) of pairs. For example, $(a, b, c) = (u, v, w)$ if and only if $a = u$, $b = v$, $c = w$. The proof of this follows from repeated use of (*).

Relations

Intuitively a relation between two mathematical objects a and b is some condition involving a and b which is either true or false for particular values of a and b. For example 'greater than' is a relation between natural numbers. Using the usual symbol $>$ we have

$$2 > 1 \quad \text{is true}$$

$$1 > 2 \quad \text{is false}$$

$$3 > 17 \quad \text{is false}$$

and so on. The relation is some sort of property of the pairs of elements a, b. In fact we must use the *ordered* pair (a, b), since for instance $2 > 1$ but not $1 > 2$.

If we know for which ordered pairs (a, b) that $a > b$ is true, then, to all intents and purposes, we have specified exactly what we mean by the relation 'greater than'. In other words, a relation may be defined by using a set of ordered pairs. Making the usual set-theorist's leap into the void:

DEFINITION. Let A and B be sets. A *relation between A and B* is a subset of $A \times B$.

If $A = B$ we talk of a relation† *on A*, which is a subset of $A \times A$.

This definition requires elucidation. For example the relation 'greater than' on $\mathcal{N}$ is the set of all ordered pairs (a, b) where $a, b \in \mathcal{N}$ and (in the usual sense) $a > b$. We might illustrate this set as follows:

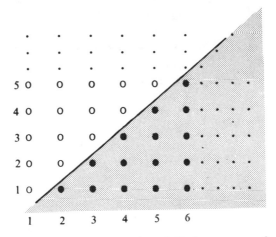

If R is a relation between sets A and B, then we say that $a \in A$ and $b \in B$ are *related* by R if $(a, b) \in R$. More commonly we shall use the notation

$$a \; R \; b \tag{*}$$

to mean $(a, b) \in R$. Then $(a, b) \notin R$ will be written $A \; \cancel{R} \; B$. This allows a piece of sleight of hand. If we denote the relation 'greater than' by the usual symbol $>$ then, letting R be $>$, we find that $a > b$ (in the sense of (*)) means the same as $(a, b) \in >$, and this by definition means that $a > b$ in the usual sense. On the other hand if $(a, b) \notin >$ we write $a \not> b$, which again corresponds to normal usage. Thus we 'recover' the standard symbolism by an unscrupulous trick of notation. This is an excellent idea—at least, mathematicians seem pleased by it—and in future we shall use the $a \; R \; b$ notation.

---

† In fact there is no need to make this distinction from a theoretical point of view, since a relation *between A and B* may be thought of as a relation *on $A \cup B$*. (Think about it!)

We consider more examples.
The relation $\geqslant$ on $\mathscr{N}$:

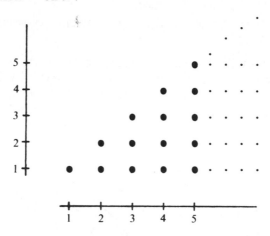

The relation $=$ on $\mathscr{N}$:

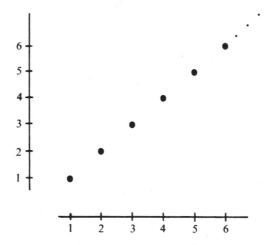

In fact, the relation $=$ on $\mathscr{N}$ is the set $\{(x, x) | x \in N\}$.

For a final example, let $X = \{1, 2, 3, 4, 5, 6\}$ and let '|' be the relation 'is a divisor of', so that $a|b$ means 'a is a divisor of b'. As a set of ordered pairs,

$$| = \{(1, \ 1), \ (1, \ 2), \ (1, \ 3), \ (1, 4), (1, 5), (1, 6), (2, 2), (2, 4),$$
$$(2, 6), (3, 3), (3, 6), (4, 4), (5, 5), (6, 6)\}.$$

In pictures:

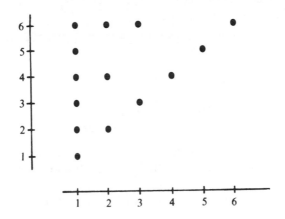

Given a relation R between sets A and B, and subset A', B' of A and B respectively, we can define a relation R' between A' and B' by

$$R' = \{(a, b) \in R \,|\, a \in A' \text{ and } b \in B'\}.$$

In fact, set-theoretically,

$$R' = R \cap (A' \times B').$$

We call R' the *restriction* of R to A' and B'. As far as the elements of A' and B' go the relations R and R' say the same thing. The only difference is that R' says nothing about elements not in A' and B'.

Equivalence relations

In both elementary and advanced mathematics the distinction between odd and even integers is often important. (To make this absolutely clear: the odd integers are those of the form $2n+1$ for integer n, namely $\ldots -5, -3, -1, 1, 3, 5, \ldots$ and the even integers are those of the form $2n$, namely $\ldots -4, -2, 0, 2, 4, \ldots$.) The set $\mathscr{Z}$ of all integers is split into two disjoint subsets

$$\mathscr{Z}_{\text{odd}} = \text{all odd integers}$$

$$\mathscr{Z}_{\text{even}} = \text{all even integers},$$

so that

$$\mathscr{Z}_{\text{odd}} \cap \mathscr{Z}_{\text{even}} = \varnothing, \; \mathscr{Z}_{\text{odd}} \cup \mathscr{Z}_{\text{even}} = \mathscr{Z}.$$

There is another way to split $\mathscr{Z}$ into these two pieces, using a *relation* which for the moment we call by the non-committal name '$\sim$'. Define for $m, n \in \mathscr{Z}$,

$$m \sim n \text{ if and only if } m - n \text{ is a multiple of 2.}$$

Then

all even integers are related by $\sim$,
all odd integers are related by $\sim$,
no even integer is related to an odd integer,
no odd integer is related to an even integer.

The fact that this is possible is a consequence of some general properties of $\sim$, and we shall analyse the situation in general to see what is required.

Imagine a set X broken up into a number of disjoint pieces.

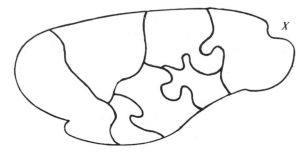

We can define a relation $\sim$ by

$$x \sim y \text{ if and only if } x \text{ and } y \text{ are both in the same piece.}$$

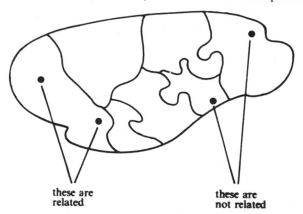

these are   these are
related   not related

Conversely we can reconstruct the pieces from the relation $\sim$: the piece to which $x \in X$ belongs is

$$E_x = \{y \in X \mid x \sim y\}.$$

If we try this with a different relation $\sim$, all sorts of things can go wrong. In particular we may not get *disjoint* pieces. For example,† if $\sim$ is the relation | on $\{1, 2, 3, 4, 5, 6\}$, then we find

$$E_1 = \{1, 2, 3, 4, 5, 6\}$$
$$E_2 = \{2, 4, 6\}$$
$$E_3 = \{3, 6\}$$
$$E_4 = \{4\}$$
$$E_5 = \{5\}$$
$$E_6 = \{6\},$$

So the set splits up according to

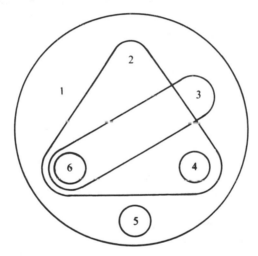

If instead we use the relation $>$ on $\mathcal{N}$, we do not even get $x \in E_x$, so E_x is in no sense 'the piece to which x belongs'.

What is it that makes the original $\sim$ work, whereas the others go wrong? We must take account of three very trite statements:

x belongs to the same piece as x,

if x belongs to the same piece as y then y belongs to the same piece as x,

† $a \mid b$ means 'a divides b'.

if x belongs to the same piece as y and y belongs to the same piece as z, then x belongs to the same piece as z.

Clearly any relation $\sim$ with the property that $x \sim y$ if and only if x and y belong to the same piece must have the three corresponding properties, which we formalize as (1), (2), (3) of the next definition.

DEFINITION.   A relation $\sim$ on a set X is an *equivalence relation* if it has the following properties for all x, y, $z \in X$.

(E1) $x \sim x$ ($\sim$ is *reflexive*),
(E2) If $x \sim y$ then $y \sim x$ ($\sim$ is *symmetric*),
(E3) If $x \sim y$ and $y \sim z$ then $x \sim z$ ($\sim$ is *transitive*).

Thus as long as we break X into *disjoint* pieces then the relation 'is in the same piece as' is an equivalence relation. We will now show that *every* equivalence relation arises in this way from a suitable choice of pieces; in fact there is an intimate connection between the two concepts. First we need a formal definition of 'breaking into disjoint pieces'.

DEFINITION.   A *partition* of a set X is a set $\mathscr{X}$ whose members are non-empty subsets of X, subject to the conditions

(P1) Each $x \in X$ belongs to some $Y \in \mathscr{X}$,
(P2) If X, $Y \in \mathscr{X}$ and $X \neq Y$, then $X \cap Y = \varnothing$.

The elements of $\mathscr{X}$ are our 'pieces'. Condition (P1) says that X is the union of all the pieces, so that each element of X lies in some piece; (P2) says that distinct pieces don't overlap. It follows that no element of X can belong to two distinct pieces.

Given an equivalence relation $\sim$ on X we define the *equivalence class* (with respect to $\sim$) of $x \in X$ to be the set

$$E_x = \{y \in X | x \sim y\}.$$

THEOREM 4.   Let $\sim$ be an equivalence relation on a set X. Then $\{E_x | x \in X\}$ is a partition of X. The relation 'belongs to the same piece as' is the same as $\sim$.

Conversely if $\mathscr{X}$ is a partition of X, let $\sim$ be defined by $x \sim y$ if and only if x and y lie in the same piece. Then $\sim$ is an equivalence relation, and the corresponding partition into equivalence classes is the same as $\mathscr{X}$.

REMARK. This theorem allows us to pass at will from an equivalence relation to a partition or back again, by a procedure which, when done twice, leads back to where we started.

PROOF.   Since $x \in E_x$ it follows that condition (P1) for a partition is satisfied. To verify (P2), suppose $E_x \cap E_y \neq \emptyset$. Then we can find $z \in E_x \cap E_y$. Then $x \sim z$ and $y \sim z$. By symmetry $z \sim y$, and then transitivity implies $x \sim y$.

We show that this implies $E_x = E_y$. For if $u \in E_x$ then $x \sim u$ and $y \sim x$, so $y \sim u$; hence $E_x \subseteq E_y$. Similarly $E_y \subseteq E_x$. This shows that $E_x = E_y$.

Thus we have proved that $E_x \cap E_y = \emptyset$ or $E_x = E_y$. But this statement is logically equivalent to (P2).

Now define $x \approx y$ to mean 'x and y are in the same equivalence class'. Then

$$x \approx y \text{ if and only if } x, y \in E_z \quad \text{for some } z$$
$$\text{if and only if } z \sim x \quad \text{and} \quad z \sim y \quad \text{for some } z$$
$$\text{if and only if } x \sim y.$$

Hence $\approx$ and $\sim$ are the same.

The second part of the theorem is proved in a similar manner, but is easier. We omit it. $\square$

An example: arithmetic modulo n

We will use the equivalence relation concept to generalize the distinction between odd and even integers and set up what is often called (in schools) 'modular arithmetic' or (in universities) 'the integers mod n'.

To begin with we specialize to $n = 3$. We define the relation $\equiv_3$ of *congruence modulo 3* on $\mathcal{Z}$ by

$$m \equiv_3 n \text{ if and only if } m - n \text{ is a multiple of 3.}$$

PROPOSITION 5. $\equiv_3$ is an equivalence relation on $\mathcal{Z}$.

PROOF.

(E1) $m - m = 0 = 3 \cdot 0$.
(E2) If $m - n = 3k$ then $n - m = 3(-k)$.
(E3) If $m - n = 3k$, $n - p = 3l$, then $m - p = 3(k + l)$. $\square$

We know that the equivalence classes (known as *congruence classes mod 3*) partition $\mathcal{Z}$. What are they? It is easiest to see this with the help of examples.

$E_0 = \{y \mid 0 \equiv_3 y\}$
  $= \{y \mid y - 0 \text{ is a multiple of } 3\}$
  $= \{y \mid y = 3k \text{ for some } k \in \mathcal{Z}\}.$

$E_1 = \{y \mid 1 \equiv_3 y\}$
  $= \{y \mid y - 1 = 3k\}$
  $= \{y \mid y = 3k + 1 \text{ for some } k \in \mathcal{Z}\}.$

$E_2 = \{y \mid 2 \equiv_3 y\}$
  $= \{y \mid y = 3k + 2 \text{ for some } k \in \mathcal{Z}\}.$

$E_3 = \{y \mid y = 3k + 3 \text{ for some } k \in \mathcal{Z}\}.$

However, $3k + 3 = 3(k + 1)$, so $E_3 = E_0$. Similarly $E_4 = E_1$, $E_5 = E_2$, $E_{-1} = E_2$, $E_{-2} = E_1$, and so on. Every integer is either of the form $3k$, $3k + 1$, or $3k + 2$ (according as it leaves remainder 0, 1, or 2 on division by 3) so we get exactly three equivalence classes:

$$E_0 = \{\ldots -9, -6, -3, 0, 3, 6, 9, \ldots\}$$
$$E_1 = \{\ldots -8, -5, -2, 1, 4, 7, 10, \ldots\}$$
$$E_2 = \{\ldots -7, -4, -1, 2, 5, 8, 11, \ldots\}.$$

So much for the equivalence relation. More intriguing is the possibility of doing *arithmetic* with these equivalence classes.

To make the notation more transparent in general, we shall change from E_n as the equivalence class of n, and instead write n_3. Hence the three classes above 0_3, 1_3, and 2_3. We put $\mathcal{Z}_3 = \{0_3, 1_3, 2_3\}$. We define operations of addition and multiplication on $\mathcal{Z}_3$ by

$$m_3 + n_3 = (m + n)_3, \tag{†}$$

$$m_3 n_3 = (mn)_3.$$

For example, $1_3 + 2_3 = 3_3 = 0_3$; $2_3 2_3 = 4_3 = 1_3$.

This may look pointless: such an impression is erroneous as will soon be seen. It may also look harmless: certain subtleties must be noticed before worrying that something may go wrong, and a little hard thinking put in to see that, after all, it doesn't.

Here is the subtle problem: the *same* class has several different names; thus $1_3 = 4_3 = 7_3 = \ldots$, $2_3 = 5_3 = 8_3 = \ldots$. The definitions (†) might (0 wicked spite!) give different answers to the same question depending on which names we use. Thus we have seen that $1_3 + 2_3 = 0_3$. But since $1_3 = 7_3$, $2_3 = 8_3$, we also have $1_3 + 2_3 = 7_3 + 8_3 = 15_3$. By a stroke of good fortune, $15_3 = 0_3$, and we can breathe again. What happens in general? If $i_3 = i'_3$ then $i - i' = 3k$ for some k: and if $j_3 = j'_3$ then $j - j' = 3l$ for some l. Now the rule (†) gives two possible answers:

$$i_3 + j_3 = (i+j)_3, \qquad i'_3 + j'_3 = (i'+j')_3.$$

However,

$$i + j = i' + 3k + j' + 3l = (i'+j') + 3(k+l),$$

so

$$(i+j)_3 = (i'+j')_3.$$

Hence we get the same answer both ways, and (†) makes sense as a definition of addition.

Similarly we must check unambiguity for multiplication. With i, j, i', j', as above, we have

$$i_3 j_3 = (ij)_3, \qquad i'_3 j'_3 = (i'j')_3.$$

But

$$ij = (i'+3k)(j'+3l) = i'j' + 3(i'l + j'k + 3kl),$$

so

$$(ij)_3 = (i'j')_3,$$

which is what we want.

This problem, much misunderstood, always arises when we try to define operations on sets by a rule of the type 'select elements from the sets, operate on these, then find the set to which the result belongs'. When, as here, the notation conceals such a process, one must be very careful to think what the notation *means* rather than just to manipulate symbols blindly. We *must* check that different choices give the same answer.

In case the reader should imagine that such checks can be dispensed with on the grounds that everything nice will work, let us consider the problem of defining exponentials in $\mathcal{Z}_3$. The natural way to do this is to mimic (†) and define

$$m_3^{n_3} = (m^n)_3.$$

For example, $2_3^{2_3} = (2^2)_3 = 4_3 = 1_3$. Using this 'definition' we can even prove theorems about the laws of exponentiation, for example

$$m_3^{n_3 + p_3} = (m^{n+p})_3 = (m^n m^p)_3 = (m^n)_3 (m^p)_3 = m_3^{n_3} m_3^{p_3}. \qquad (\dagger\dagger)$$

However, we would be living in a fools' paradise. For, since $2_3 = 5_3$, the rule (*) also tells us that

$$2_3^{2_3} = 2_3^{5_3} = (2^5)_3 = (32)_3 = 2_3.$$

Since $1_3 \neq 2_3$ this shows that (†) is nonsense—but clever and plausible nonsense, the most dangerous kind.

In common parlance, what we must check is that the operations are 'well defined'. Really this is over-polite: what we are checking is that they are 'defined' at all! A fraudulent definition is no true definition whatsoever.

Having digressed at length, let us return to the arithmetic of $\mathscr{Z}_3$. We can write out addition and multiplication tables:

+	0_3	1_3	2_3
0_3	0_3	1_3	2_3
1_3	1_3	2_3	0_3
2_3	2_3	0_3	1_3

×	0_3	1_3	2_3
0_3	0_3	0_3	0_3
1_3	0_3	1_3	2_3
2_3	0_3	2_3	1_3

It can be verified that many of the usual laws of arithmetic hold (such as $x + y = y + x$, $x(y + z) = xy + xz$) although there are some surprises such as

$$((1_3 + 1_3) + 1_3 + 1_3) = 1_3.$$

Instead of 3, we can use any integer n and do arithmetic modulo n. We define a relation $\equiv_n$ on $\mathscr{Z}$ by

$$x \equiv_n y \text{ if and only if } x - y \text{ is a multiple of } n.$$

We get n distinct equivalence classes $0_n, 1_n, 2_n, \ldots, (n-1)_n$; while $n_n = 0_n$, $(n+1)_n = 1_n$, etc.; and x_n consists of the integers which leave remainder x on division by n. The set $\mathscr{Z}_n$ of equivalence classes admits operations of arithmetic defined in the same way as (†).

We shall discuss these ideas further in chapter 10.

Order relations

The various order relations usually encountered in dealing with numbers can be exemplified by the statements $4 < 5$, $7 > 2\pi$, $x^2 \geq 0$, $1 - x^2 \leq 1$ for any real number x. Fortunately the relations $<$, $>$, $\leq$,

$\geqslant$ are all connected with each other:

$x < y$ means the same as $y > x$,

$x \leqslant y$ .... $y \geqslant x$,

$x \leqslant y$ .... $x < y$ or $x = y$,

$x < y$ .... $x \leqslant y$ and $x \neq y$.

This means that we only need study one of them and translate the results to the others. In handling actual numbers, it might be preferable to consider one of the strict relations $<$ or $>$. In general we wouldn't write $2 + 2 \geqslant 4$, simply because we know something more precise, $2 + 2$ *equals* 4. Likewise we would normally write $2 + 2 > 3$, because that contains more exact information than $2 + 2 \geqslant 3$. In passing to general statements, the situation changes somewhat. For instance, it is true that if $a_n \to a$, $b_n \to b$ and $a_n \geqslant b_n$, then $a \geqslant b$, but $a_n > b_n$ does not imply $a > b$. (Counterexample is given by $a_n = 1/n$, $b_n = 0$.) Here there is a slight preference towards the weak inequalities $\leqslant$, $\geqslant$. We will begin with the latter.

DEFINITION.   A relation R on a set A is said to be a *weak order relation* if

(WO 1) $a \ R \ b$ and $b \ R \ c$ imply $a \ R \ c$,

(WO 2) Either $a \ R \ b$ or $b \ R \ a$ (or both)

(WO 3) $a \ R \ b$ and $b \ R \ a$ imply $a = b$.

These properties evidently hold for both of the relations $\leqslant$, $\geqslant$ on the set of real numbers, which may seem rather strange since one means "bigger" and the other "smaller". But looking at the real numbers as points on a line, we see that, by a reflection, we can turn the order round, interchanging left and right, and this interchanges $\leqslant$ and $\geqslant$. It's only when we start doing arithmetic, and require the fact that $a \geqslant 0$, $b \geqslant 0$ implies $ab \geqslant 0$, do we find a property of $\geqslant$ which does not hold for $\leqslant$. We will postpone this consideration until we study arithmetic in chapter 9. Meanwhile it is sufficient to see that weak order relations come in pairs. Given such a relation R, we can define R', the reverse of R by

$$a \ R' \ b \quad \text{means} \ b \ R \ a. \tag{*}$$

The reverse R' is also a weak order relation and reversing once again we come back to $R'' = R$.

EXAMPLE 1.   If $A = \{a, b, c\}$ where a, b, c are all distinct, then a weak order on A can be defined by $a \mathrel{R} b, a \mathrel{R} c, b \mathrel{R} c, a \mathrel{R} a,$ $b \mathrel{R} b, c \mathrel{R} c.$ We can visualize this by considering a, b, c in a row, with $x \mathrel{R} y$ meaning x is to the left of y or $x = y$.

$$a \quad b \quad c$$
$$\circ \quad \circ \quad \circ$$

The reverse of R simply puts the elements in order c, b, a.

EXAMPLE 2   Define the order relation R on the plane by: $(x_1, y_1) \mathrel{R} (x_2, y_2)$ means

either $y_1 = y_2$ and $x_1 \leqslant x_2$ (both together),

or $y_1 < y_2$.

This at first looks bizarre, but in a picture we see that $(x_1, y_1) \mathrel{R} (x_2, y_2)$ simply means that either (x_1, y_1) and (x_2, y_2) are on the same horizontal line with (x_1, y_1) to the left of (or equal to) (x_2, y_2) or (x_1, y_1) is on a horizontal line strictly below the one through (x_2, y_2).

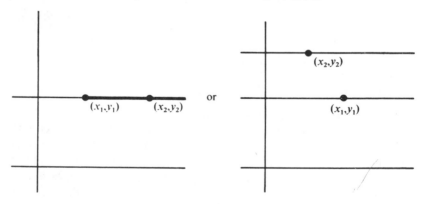

EXAMPLE 3. $A = \{\{0\}, \{0, 1\}, \{0, 1, 2\}, \{0, 1, 2, 3\}\},$

$x \mathrel{R} y$ means $x \subseteq y$ for $x, y \in A$.

Here we have $\{0\} \subseteq \{0, 1\} \subseteq \{0, 1, 2\} \subseteq \{0, 1, 2, 3\}$.

Inclusion of sets satisfies

$$X \subseteq Y \quad \text{and} \quad Y \subseteq Z \quad \text{implies } X \subseteq Z,$$
$$X \subseteq Y \quad \text{and} \quad Y \subseteq X \quad \text{implies } X = Y,$$

but for arbitrary sets X, Y we may have $X \not\subseteq Y$, and $Y \not\subseteq X$. This means that set inclusion in general satisfies (WO 1), (WO 3), but not (WO 2). A relation R on a set A satisfying (WO 1) and (WO 3) is said to be a *partial order*, and A a *partially ordered set*, or, with some loss in dignity, a poset. Given any set A of sets, then inclusion always yields a partial order.

Let R be a weak order on a set A, then the corresponding strict order S is given by:

$x \, S \, y$ means precisely $x \, R \, y$ and $x \neq y$. For example if R is $\leqslant$, then S is $<$.

PROPOSITION 6.   A strict order S on a set A satisfies:

(SO1) $a \, S \, b$ and $b \, S \, c$ imply $a \, S \, c$

(SO2) Given $a, b \in A$, then precisely one of the following hold (and not the other two):

$$a \, S \, b, b \, S \, a, a = b.$$

PROOF.   Suppose that $a \, S \, b$ and $b \, S \, c$ hold. Then $a \, R \, b$ and $b \, R \, c$, and by (WO1) we deduce $a \, R \, c$. We cannot have $a = c$, for substituting in $b \, R \, c$, we get $b \, R \, a$ and by (WO3), this and $a \, R \, b$ would give $a = b$, contradicting $a \, S \, b$. This verifies (SO1). From (WO2) we get $a, b \in A$ implies $a \, R \, b$ or $b \, R \, a$, so $a \, S \, b$ or $a = b$ or $b \, S \, a$. But no two of these can hold simultaneously because $a = b$ contradicts the definitions of both $a \, S \, b$ and $b \, S \, a$, and were $a \, S \, b$, $b \, S \, a$ to hold simultaneously, then $a \, R \, b$, $b \, R \, a$ hold, so (WO3) gives $a = b$, contradicting $a \, S \, b$ once more. This verifies (SO2) $\square$. Proposition (SO2) is usually referred to as the *trichotomy law*. (Just as a dichotomy is two mutually exclusive possibilities, a trichotomy is three, in this case $a \, S \, b, b \, S \, a$, and $a = b$.) For the strict order $<$ on the real numbers the three mutually exclusive possibilites are $a < b$, $b < a$, $a = b$.

We remarked earlier that we could return to the weak order $\leqslant$ from $<$ through the connection

$$a \leqslant b \quad \text{means precisely} \quad a < b \quad \text{or} \quad a = b.$$

The same happens for any strict order. Given a relation S on a set A satisfying (SO1) and (SO2), define

$$a \, R \, b \quad \text{to mean} \quad a \, S \, b \quad \text{or} \quad a = b.$$

It is easy to verify that R satisfies (WO1)–(WO3), and that we can pass freely from a weak order to the corresponding strict order and back again. In this manner the notions of weak and strict order are interchangeable. Although we have taken (WO1)–(WO3) as the basic axioms and proved the properties (SO1), (SO2), we could just as easily reverse their status taking (SO1), (SO2) as basic axioms and deducing (WO1)–(WO3).

Exercises

1. Write out the proofs of propositions 2(b), 2(c), and 2(d).

2. Prove that

$$(\cup S) \times (\cup T) \subseteq \cup \{X \times Y \mid X \in S, \, Y \in T\}$$
$$(\cap S) \times (\cap T) = \cap \{X \times Y \mid X \in S, \, Y \in T\}$$

for all sets S, T of sets. Show that in the first formula we cannot replace '$\subseteq$' by '$=$'.

3. If $A = \varnothing$, show that $A \times B = \varnothing = B \times A$ for every set B. When $A \neq \varnothing$, show that $A \times B = A \times C$ implies $B = C$. Given $A \times B = B \times A$, what can be deduced about A and B?

4. Let $A = \mathcal{N} \times \mathcal{N}$. Define the relation R on A by

$$(m, n) \, R \, (r, s) \quad \text{means } m + s = r + n.$$

Show that R is an equivalence relation.

If $B = \{(x, y) \in \mathcal{X} \times \mathcal{X} \mid y \neq 0\}$, and S is the relation on B given by

$$(a, b) \, S(c, d) \text{ if and only if } ad = bc,$$

is S an equivalence relation? Prove your assertion.

5. How many distinct equivalence relations exist on $\{1, 2, 3, 4\}$?

6. Recall the properties (E1), (E2), (E3) for an equivalence relation. Which of these properties is satisfied by the relation between $x, y \in R$ given by:

(a) $x < y$
(b) $x \geq y$
(c) $|x - y| < 1$
(d) $|x - y| \leq 0$
(e) $x - y$ is rational
(f) $x - y$ is irrational
(g) $(x - y)^2 < 0$.

7. Is there a mistake in the following proof that (E2) and (E3) imply (E1)? If so, what is it?

Let $a \sim b$. By (E2), $b \sim a$. By (E3), if $a \sim b$ and $b \sim a$, then $a \sim a$. This proves (E1).

8. Recall that an equivalence relation is defined to satisfy axioms (E1), (E2), (E3). Gives examples of relations (the more elegant, the better) satisfying

  (a) None of the properties (E1), (E2), (E3),
  (b) (E1) but not (E2) or (E3),
  (c) (E2) but not (E1) or (E3),
  (d) (E3) but not (E1) or (E2),
  (e) (E2) and (E3) but not (E1),
  (f) (E1) and (E3) but not (E2),
  (g) (E1) and (E2) but not (E3).

9. Write out addition and multiplication tables for the integers mod 4, mod 5, and mod 6.
Find all $a, b \in \mathscr{Z}_{12}$ such that $ab = 0_{12}$.

10. Define a relation R on $\mathscr{N}$ by

$$a \ R \ b \text{ means } a \text{ divides } b,$$

that is, $b = ac$ for some $x \in \mathscr{N}$.
Is R an order relation? If so, is it a weak order or a strict order?

11. Let $X = \{1, 2, 6, 30, 210\}$ and define a relation S on X by

$$a \ S \ b \quad \text{means} \quad a \text{ divides } b.$$

Is S an order relation? If so, is it a weak order or a strict order?

12. Let A be a set with a (strict) order relation S and B a set with (strict) order relation T. Define the lexicographic relation L on $A \times B$ by

$$(a, b) \ L \ (c, d) \quad \text{means: either } a \ S \ c$$
$$\text{or } a = c \text{ and } b \ T \ d.$$

Is this an order relation? What is the connection between this and a dictionary?

5

Functions

THE concept of a 'function' is of enormous importance throughout the whole of modern mathematics, at all levels. Traditionally the function concept first became prominent in the calculus, which is *about* functions: how to differentiate them or integrate them.

Early attempts to develop a concept of functions *in general* were somewhat confused and unsatisfactory, largely because they tried to do too much at once. The function concept as it is now understood evolved gradually from these attempts: it has great generality and great simplicity. So general is the function concept that, when doing calculus, extra conditions must first be imposed to restrict the class of functions to those which *can* be differentiated or integrated. Thus the desired object is achieved by taking a very general definition of 'function' and then selecting more special types of function by imposing more conditions.

In this chapter it is the general concept which we wish to consider—after all, this isn't a book on calculus! We shall develop the general concept gradually out of a discussion of familiar examples and then analyse the consequences. We shall discuss some of the more general properties that functions can have. The graph of a function will be introduced and related both to the formal definition and the traditional picture.

Some traditional functions

Traditionally one introduces a 'variable' x, which is usually supposed to be a real number, and talks of a 'function $f(x)$ of x'. The main point is that we are able to work out the value of $f(x)$ for any given x (possibly under restrictions such as $x \neq 0$, $x > 0$, depending on the function involved). Thus the exponential function takes value e^x for any real number x (where $e = 2 \cdot 71828 \ldots$); the sine, cosine, or tangent functions take values $\sin(x)$, $\cos(x)$, $\tan(x)$, for all real x (except that for the tangent we have to assume x is not an odd integer multiple of $\pi/2$ in order for the definition $\tan(x) = \sin(x)/\cos(x)$ to make sense); the logarithmic function takes value $\log(x)$ when x is real and $x > 0$; the reciprocal function takes value $1/x$ for real $x \neq 0$; the square function takes value x^2 for any real x. Other functions, such as the factorial function $x!$, are defined only for x a positive integer.

What do all of these examples have in common? It is our ability, in principle, to calculate the value of the function corresponding to the relevant values of x. In other words, a function associates to each relevant real number x a value $f(x)$ which is also a real number. In the above examples we have, respectively, $f(x) = e^x$, $\sin(x)$, $\cos(x)$ $\tan(x)$, $\log(x)$, $1/x$, x^2, or $x!$.

We should not confuse the *values* of the function with the function itself. It is not $\log(x)$ that is the function: it is the rule 'take the logarithm of' which allows us to *work out* the value. It is, in a sense, the symbol 'log'. Thus we think of a function f as some 'rule' which, for any real number x (perhaps subject to restrictions), defines another real number $f(x)$. We shall insist that $f(x)$ be uniquely defined: a 'rule' which gives two different answers to the same question is not especially useful. This means we must be careful with such functions as 'square root', specifying whether we mean the positive square root or the negative one. Don't worry about this now, we'll return to it later when our ideas are better established.

The general function concept

The most general definition of a function comes from the traditional one by relaxing the requirement that x and $f(x)$ be real numbers. (Even in traditional mathematics one would certainly want to allow complex numbers, and in fact a wide variety of non-numerical things as well: the easiest and most satisfactory thing is not to impose restrictions of any kind on the nature of x or $f(x)$). Then we have to be more precise regarding what we mean by 'rule'.

We take two sets A and B. As a preliminary definition: a *function f from A to B* is a rule which assigns to each $a \in A$ a *unique* element $f(a) \in B$.

This is a very wide definition. It includes all of the aforementioned examples: we take A to be a suitable subset of $\mathscr{R}$, and B to be $\mathscr{R}$. Thus for the exponential function we have $A = \mathscr{R}$, $B = \mathscr{R}$, and the rule is that $f(x) = e^x$. For the logarithm $A = \{x \in \mathscr{R} \mid x > 0\}$, $B = \mathscr{R}$, and $f(x) = \log(x)$. For the reciprocal, $A = \{x \in \mathscr{R} \mid x \neq 0\}$, $B = \mathscr{R}$, and $f(x) = 1/x$. For the factorial we have $A = \mathscr{N}$, $B = \mathscr{N}$, and $f(x) = x!$.

Examples of rather different types of functions which this definition allows include:

$A = \{\text{all circles in the plane}\}$, $B = \mathcal{R}$, $f(x) = $ the radius of x.

$A = \{\text{all circles in the plane}\}$, $B = \mathcal{R}$, $f(x) = $ the area of x.

$A = \{\text{all subsets of } \{0, 2, 4\}\}$, $B = \mathcal{N}$, $f(x) = $ the smallest element of x.

$A = \{\text{all subsets of } \{0, 1, 2, 3, 4, 5, 6, 7\}\}$, $B = \{0, 1, 2, 3, 4, 5, 6, 7, 8\}$, $f(x) = $ the number of elements of x.

$A = \mathcal{Z}$, $B = \{0, 1, 2\}$, $f(x) = $ the remainder on dividing x by 3.

$A = \{\text{camel, lion, elephant}\}$, $B = \{\text{January, March}\}$, $f(\text{camel}) = $ March, $f(\text{lion}) = \text{January}$, $f(\text{elephant}) = \text{March}$.†

The following terminology is standard. We call A the *domain* of f and B the *codomain*, and write

$$f: A \to B$$

to mean 'f is a function with domain A and codomain B'.

The main item still on the agenda is that troublesome 'rule'. We obtain a formal definition in exactly the same sort of way that we obtain one for 'relation', by judicious use of ordered pairs. We want to associate to each $x \in A$ an element $f(x) \in B$. One way to do this is to stick them together in an ordered pair $(x, f(x))$. The 'rule' is then the entire set of ordered pairs $(x, f(x))$ as x runs through A, and this is of course a subset of $A \times B$. The requirement that $f(x)$ is defined for every $x \in A$ translates as the requirement that for any $x \in A$ there is *some* element (x, y) in the set: the uniqueness of $f(x)$ translates as a requirement that such a y should be *unique*. Now we can see how the set of pairs captures the rule: to find $f(x)$ look in the set for a pair (x, y): this exists and is unique, so we put $f(x) = y$.

Formally, then:

DEFINITION. Let A and B be sets. A *function* $f: A \to B$ is a subset f of $A \times B$ such that

(F1) If $x \in A$ there exists $y \in B$ such that $(x, y) \in f$.

(F2) Such an element y is unique: in other words, if $x \in A$ and $y, z \in B$ are such that $(x, y) \in f$ and $(x, z) \in f$, it follows that $y = z$.

A function is also called a *map* or *mapping*.

---

† This is of course a pretty silly sort of function, and not one of mathematical importance. It illustrates that quite arbitrary definitions of $f(x)$ may be made. Actually, this one isn't quite so arbitrary as it may seem. A certain zoo has three main animal-houses: the camel-house, the lion-house, and the elephant-house. Once a year the houses are redecorated: the lion-house in January, the others in March. Now $f(x) = $ the month in which the x-house is redecorated.

In terms of this definition the 'square' function defined on $\mathscr{R}$ is the subset of $\mathscr{R} \times \mathscr{R}$ given by

$$\{(x, x^2) \mid x \in \mathscr{R}\}.$$

The curious function above is the set

$$\{(\text{camel, March}), (\text{lion, January}), (\text{elephant, March})\}.$$

We recover the usual notation by defining $f(x)$ to be the unique element $y \in B$ such that $(x, y) \in f$.

The definition by ordered pairs is formally very nice, because it states everything in terms of sets. But it is pedantic and pointless to use ordered pairs when we wish to define a specific function. We will use a form of words along the following lines:

'Define a function $f: A \to B$ by $f(x) = \ldots$ for all $x \in A$'.

In any particular case the dots ... will be replaced by a specific prescription to find $f(x)$ given x. This is to be interpreted formally as:

'f is the subset of $A \times B$ consisting of all pairs $(x, f(x))$ for $x \in A$'.

Then all we must be careful of is to check that the prescription defines $f(x)$ uniquely, and that $f(x) \in B$, for all $x \in A$.

Just to clarify the point, here's an example. Define the function $f: \mathscr{N} \to \mathscr{Q}$ by

$$f(n) = \sqrt{2} \text{ to } n \text{ decimal places.}$$

Then

$$f(0) = 1$$
$$f(1) = 1.4$$
$$f(2) = 1.41$$
$$f(3) = 1.414$$

etc.

The formal set of ordered pairs is the set of all

$$(n, \sqrt{2} \text{ to } n \text{ decimal places}),$$

namely

$$\{(1, 1.4), (2, 1.41), (3, 1.414), \ldots\}.$$

The advantage of the informal usage is manifest. But the fact that we know how to translate it into formal terms means that the informality is safe.

To close this section, here are a number of statements that look as if they define functions, but on closer inspection fail in some respect.

$$\text{'Define } f \colon \mathcal{R} \to \mathcal{R} \text{ by } f(x) = \frac{x^2 + 17x + 93}{x + 1}.$$

This does not define a function since when $x = -1$, $1/(x+1)$ is not defined, so $f(-1)$ has not been specified as a real number. If we change the definition to start '$f \colon \mathcal{R}\setminus\{-1\} \to \mathcal{R} \ldots$' then we're all right.

'Define $f \colon \mathcal{Q} \to \mathcal{Q}$ by $f(x) = \sqrt{x}$ (positive square root).' This doesn't define a function because for some x, e.g. $x = 2$, the value $f(x) = \sqrt{2}$ does not belong to $\mathcal{Q}$. If we change the second $\mathcal{Q}$ to an $\mathcal{R}$ then all will be well.

'Define $f \colon \mathcal{R} \to \mathcal{Q}$ by $f(x) = $ the rational number nearest to x'.

This does not define a function: the supposed $f(x)$ does not exist.

'Define $f \colon \mathcal{R} \to \mathcal{Z}$ by $f(x) = $ the integer nearest to x.'

This almost works: the trouble is that both 0 and 1 are equidistant from $\frac{1}{2}$ so $f(\frac{1}{2})$ is not defined *uniquely*.

General properties of functions

If $f \colon A \to B$ is a function, then an important subset of B is the *image* of f, defined to be

$$f(A) = \{f(x) \mid x \in A\}.$$

The image of f is the set of values obtained by working out $f(x)$ for all x in the domain. It need not be the whole codomain; for example if $f \colon \mathcal{R} \to \mathcal{R}$ has $f(x) = x^2$ then the image is the set of positive reals, and is not the whole codomain $\mathcal{R}$.

The lack of symmetry in the definition of a function may seem disturbing. We ask that $f(x)$ be defined for all $x \in A$, yet we do not ask that every $b \in B$ be of the form $f(x)$. The reason for this is a pragmatic one. When we *use* a function we want to be sure that it is defined; knowledge of the precise domain is thus essential. It is less crucial to know exactly where the values $f(x)$ lie, and we choose the codomain to be whatever is convenient. For instance, if we define

$$f \colon \mathcal{N} \to \mathcal{R}$$

by

$$f(n) = \sqrt[3]{n!}$$

then the image of f is the set of cube roots of factorials

$$\{1, \sqrt[3]{2}, \sqrt[3]{6}, \sqrt[3]{24}, \sqrt[3]{120}, \ldots\}$$

which is not a very nice set. Images in general can be pretty revolting. So we tend to define a function in terms of a codomain, and to leave aside the calculation of exactly which part of the codomain we really require, in the hope (often fulfilled) that it is not needed. If it is, we can work it out.

This brings us to another minor point. Strictly speaking we cannot talk of 'the' codomain of a function. Consider

$$f: \mathcal{R} \to \mathcal{R} \qquad f(x) = x^2,$$

$$g: \mathcal{R} \to \mathcal{R}^+ \qquad g(x) = x^2,$$

where $\mathcal{R}^+ = \{x \in \mathcal{R} \,|\, x \geq 0\}$. The first has codomain $\mathcal{R}$, the second $\mathcal{R}^+$; yet the formal definition of a function as a set of ordered pairs leads to the set $\{(x, x^2) \,|\, x \in \mathcal{R}\}$ in both cases. The functions f and g are equal.

The point is that the codomain of a function is ambiguous. *Any* set which includes the range of the function will do as codomain. The domain, on the other hand, is uniquely specified.

One can get past this by being more pedantic, defining a function to be a triple (f, A, B) rather than just a set of ordered pairs f. But for us this is not worth doing: it is easier to accept ambiguity in the codomain. The notation $f: A \to B$ then tells us which of the possible codomains is intended in any particular instance.

It is often convenient to picture a function $f: \mathcal{R} \to \mathcal{R}$ by drawing its graph. For sets other than $\mathcal{R}$ it is often better to think of a function in terms of a picture like this:

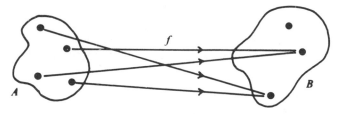

For each $x \in A$ the value of $f(x)$ is to be found at the far end of the corresponding arrow. The definition of a function, expressed in pictorial terms, then becomes:

(F1') Every element of A is at the tail end of a unique arrow,

(F2') All the arrowheads end up in B.

This is just a pictorial device, on a par with Venn diagrams, but it is useful as a source of motivation and simple examples.

On such a picture the range of f is just the set of elements of B which lie on the arrows:

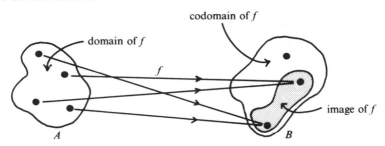

The range of f will then be the whole codomain B if every element of B is at the end of some arrow. This motivates the more formal:

DEFINITION.   A function $f: A \to B$ is a *surjection* (to B) or is *onto* B if each element of B is of the form $f(x)$ for some $x \in A$.

Whether or not a function is a surjection depends on the choice of codomain. The statement 'f is a surjection' can be made only if it is clear from the context which codomain is intended—as will be the case in a phrase such as '$f: A \to B$ is a surjection' where the codomain is B. The next examples should clarify the matter.

(1) $f: \mathcal{R} \to \mathcal{R}$, where $f(x) = x^2$. This is not a surjection to $\mathcal{R}$, since no negative real number is the square of a real number; in particular $-1 \in \mathcal{R}$ but is not of the form x^2 for any $x \in \mathcal{R}$.

(2) $f: \mathcal{R} \to \mathcal{R}^+$, where $f(x) = x^2$. This is a surjection to $\mathcal{R}^+$, since every positive real number has a square root which is real.

(3) $f: A \to B$ where $A = \{$all circles in the plane$\}$, $B = \mathcal{R}^+$, and $f(x) = $ the radius of x. This is a surjection to $\mathcal{R}^+$, since given any positive real number we can find a circle with that number as radius.

If no element of B lies at the end of two different arrows, we have another important type of function:

DEFINITION: A function $f: A \to B$ is an *injection*, or is *one–one*, if for all $x, y \in A$, $f(x) = f(y)$ implies $x = y$.

This time the precise choice of codomain does not lead to any problems: if f is an injection with one choice of codomain it is also an

injection with any other choice. Here are some examples:
(1) $f: \mathcal{R} \to \mathcal{R}$, where $f(x) = x^2$. This is not an injection, since $f(1) = 1 = f(-1)$ but $1 \neq -1$.
(2) $f: \mathcal{R}^+ \to \mathcal{R}$, where $f(x) = x^2$. This is an injection: if x and y are positive reals and $x^2 = y^2$ then $0 = x^2 - y^2 = (x-y)(x+y)$, so either $x - y = 0$ and $x = y$, or $x + y = 0$ which is impossible with both x and y positive unless $x = y = 0$. Either way, $x = y$.
(3) $f: \mathcal{R} \backslash \{0\} \to \mathcal{R}$, where $f(x) = 1/x$. This is an injection, since if $1/x = 1/y$ then $x = y$.

Nicest of all are functions with both of these properties:

DEFINITION.   A function $f: A \to B$ is a *bijection*, or is a *one-to-one correspondence*, if it is both an injection and a surjection (to B).

Again, this depends on the choice of codomain. The terms 'one–one' and 'one-to-one correspondence' are often confused, and we shall avoid them both. Instead of saying 'f is a bijection (injection, surjection)' we shall often say 'f is *bijective (injective, surjective)*'.

Clearly $f: A \to B$ is a bijection if and only if every $b \in B$ is of the form $b = f(x)$ for a *unique* $x \in A$.

All combinations of injectivity and surjectivity can occur, as the following pictures illustrate:

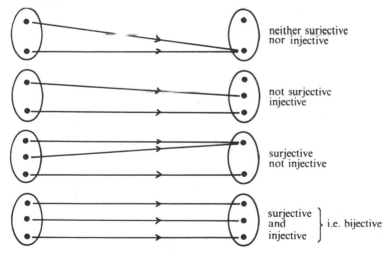

neither surjective nor injective

not surjective injective

surjective not injective

surjective and injective $\}$ i.e. bijective

One very important and very trivial function can be defined on each set A. The *identity* function $i_A: A \to A$ is defined by $i_A(a) = a$ for all $a \in A$. This is obviously a bijection.

The graph of a function

We have two competing ways of picturing functions: the graph (for functions $\mathscr{R} \to \mathscr{R}$) and the blobs-and-arrows diagram. There are interesting connections between the two. A blobs-and-arrows picture of the function $f(x) = x^2$ $(x \in \mathscr{R})$ looks something like this:

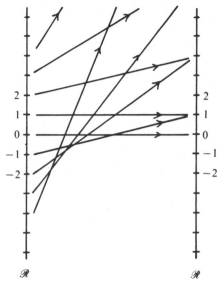

We can disentangle the arrows better if we place A horizontally and let it overlap B at 0:

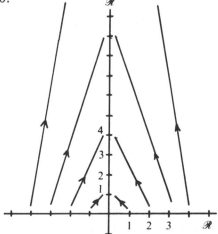

However, it is more interesting to use arrows that run only vertically or horizontally:

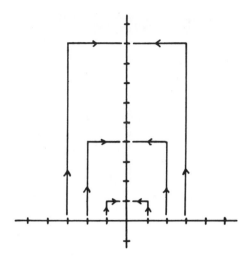

This makes it clear that the important thing is where the corner occurs. If we vary x, all the corners lie on a curve:

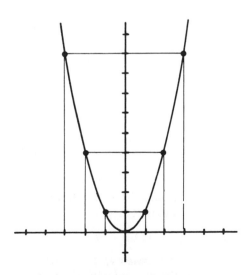

Now we can eliminate the arrows, leaving just the usual graph of f:

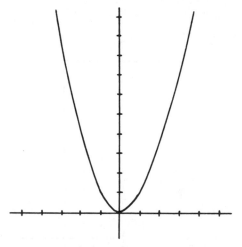

Conversely, given this graph we can put the arrows back: starting at $x \in A$ we move vertically until we hit the graph, then horizontally until we hit B. This point will be $f(x)$.

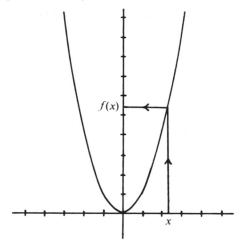

What is the graph set-theoretically? The plane is $\mathscr{R} \times \mathscr{R}$, and the corner in the arrow from x to $f(x)$ occurs at $(x, f(x))$. So the graph of f is the set

$$\{(x, f(x)) | x \in \mathscr{R}\}.$$

But this, in formal terms, is the *same* as f. By drawing $\mathscr{R} \times \mathscr{R}$ as a plane, we are led to the graph as the natural picture for f.

For a general function $f: A \to B$ we need a corresponding picture. Now we have a way to draw $A \times B$, and we use this instead of the plane. Thus the camel-lion-elephant function from two sections back has the 'graph'

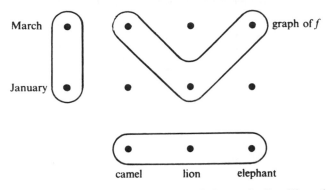

Now, we draw arrows from elements of A, vertically till we hit the graph, then horizontally till we hit B, getting:

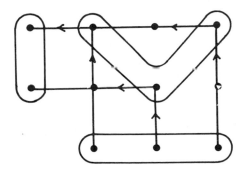

A little distortion recovers the blobs-and-arrows picture:

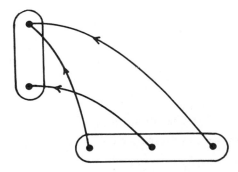

Notice that, strictly speaking, the graph picture for a function $f: \mathscr{R} \rightarrow \mathscr{R}$ should look like:

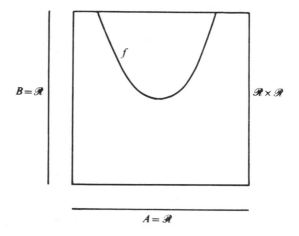

However the traditional picture, with A and B drawn *on* the plane as 'axes', is more familiar and convenient:

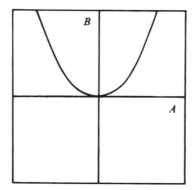

But it should be remembered that these 'axes' are not *part* of the graph, but serve the role of labels for the points (x, y) of the plane.

Composition of functions

If $f: A \rightarrow B$ and $g: C \rightarrow D$ are two functions, and if the image of f is a subset of C, then we can *compose* f and g by 'first doing f, then g'. In

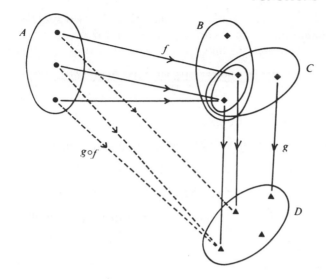

formal terms, under these hypotheses, we define a function

$$g \circ f \colon A \to D$$

by†

$$g \circ f(x) = g(f(x)).$$

Of course we must verify that $g \circ f$ is a function from A to D, but this is easy.

A very useful property of composition of functions is that it is associative, in the following sense:

PROPOSITION 1. Let $f \colon A \to B$, $g \colon C \to D$, $h \colon E \to F$ be functions, such that the image of f is a subset of C and the image of g is a subset of E. Then the two functions

$$h \circ (g \circ f) \colon A \to F$$

$$(h \circ g) \circ f \colon A \to F$$

are equal.

---

† It is a pity that $g \circ f$ corresponds to 'first f, then g', since a more natural notation would seem to be $f \circ g$. But the latter would make the definition read $f \circ g(x) = g(f(x))$, which looks wrong. One way out is to write $(x)f$ instead of $f(x)$ and let composition be given by $(x)f \circ g = ((x)f)g$. But this looks odd too!

PROOF.   By 'equal' here we mean that the two subsets of $A \times F$ which define the functions are equal; this in turn means that for each $x \in A$ the two functions take the same value. Now

$$h \circ (g \circ f)(x) = h(g \circ f(x)) = h(g(f(x)));$$
$$(h \circ g) \circ f(x) = h \circ g(f(x)) = h(g(f(x))),$$

which proves the theorem.   □

Identity functions have nice properties under composition:

PROPOSITION 2.   If $f: A \to B$ is a function, then

$$f \circ i_A = f, \qquad i_B \circ f = f.$$

PROOF.   This is a routine verification from the definitions.   □

Inverse functions

We think of a function $f: A \to B$ as something which takes $x \in A$ and *does something* to it, namely produces $f(x) \in B$. Sometimes we can find a function g which 'undoes' what f 'does', and we call this an inverse function to f. Precise analysis reveals some pitfalls.

DEFINITION.   Let $f: A \to B$ be a function. Then a function $g: B \to A$ is called a

left inverse for f if $g(f(x)) = x$ for all $x \in A$,
right inverse for f if $f(g(y)) = y$ for all $y \in B$,
inverse for f if it is both a left and a right inverse for f.

Note that the three conditions may be stated in equivalent terms:

$$g \circ f = i_A,$$
$$f \circ g = i_B,$$
$$g \circ f = i_A \text{ and } f \circ g = i_B.$$

Here are some illustrations, using single arrows $\rightarrowtail$ for f and double arrows $\twoheadrightarrow$ for g.

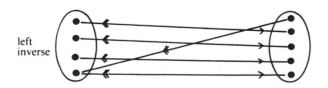

left
inverse

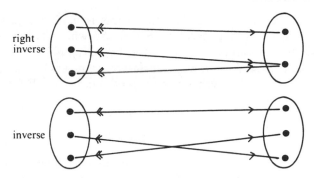

right
inverse

inverse

These pictures suggest a useful criterion:

THEOREM 3.   A function $f: A \to B$ has a:
(a) left inverse if and only if it is injective,
(b) right inverse if and only if it is surjective,
(c) inverse if and only if it is bijective.

PROOF (a).   Suppose f has a left inverse g. To prove f injective, suppose that $f(x) = f(y)$. Then $x = g(f(x)) = g(f(y)) = y$, so f is indeed an injection.

Conversely, suppose f injective. If $y \in B$ and $y = f(x)$ define $g(y) = x$. By injectivity this x is unique. If y is not an element of the range of f, no such x exists; we then pick *any* $a \in A$ and define $g(y) = a$. Now $g(y)$ is defined for all $y \in B$ and $g: B \to A$ is a function. But $g(f(x)) = x$ by the definition of g, so g is a left inverse.

(b) Suppose that f has a right inverse g. If $y \in B$ then $y = f(g(y))$, so is of the form $f(x)$ for $x = g(y)$. Hence f is a surjection to B.

Conversely suppose f surjective. Let $y \in B$. Then $y = f(x)$ for some $x \in A$, not necessarily unique. For each $y \in B$ defines $g(y)$ to be *one particular choice* of element in A such that $f(g(y)) = y$. Then g is a function and a right inverse to f.

(c)   The function f has an inverse if and only if it has a left inverse g which is also a right inverse. This implies that f is both injective and surjective, hence bijective. Now if f is bijective it has a left inverse g, and it is easy to verify that this g is also a right inverse. Hence f has an inverse. $\square$

EXAMPLES
(1) $f: \mathcal{R} \to \mathcal{R}$, $f(x) = x^3$. This is bijective, and has inverse $g: \mathcal{R} \to \mathcal{R}$, $g(x) = \sqrt[3]{x}$.

(2) $f: \mathcal{R} \to \mathcal{R}$, $f(x) = x^2$. As it stands this is neither injective nor surjective, so has neither kind of inverse. So what has happened to square roots? We can make f surjective by taking $\mathcal{R}^+ = \{x \in \mathcal{R} | x \geq 0\}$ as codomain. Now $f: \mathcal{R} \to \mathcal{R}^+$ is surjective, and $g(x) = \sqrt{x}$ (positive square root) is a *right* inverse since $f(g(x)) = (\sqrt{x})^2 = x$. But it is *not* a left inverse, since

$$\sqrt{x^2} = x \quad \text{if } x \geq 0,$$

$$\sqrt{x^2} = -x \quad \text{if } x < 0.$$

(3) We assume for this example properties of exponentials and logarithms which we have not proved rigorously in this book. Let $f: \mathcal{R} \to \mathcal{R}$ be given by $f(x) = e^x$. This is injective, for if $e^x = e^y$ then $e^{x-y} = 1$ so $x - y = 0$ so $x = y$. It has a right inverse, defined by (say)

$$g(y) = \log y \quad (y > 0)$$

$$g(y) = 273 \quad (y \leq 0).$$

For $g(f(x))$ is calculated as follows: $f(x) = e^x$ which is positive, so $g(f(x)) = g(e^x) = \log e^x = x$. The arbitrary 273 does not enter into this calculation: it is there merely to define g on the *whole* of $\mathcal{R}$. Any other definition would do for negative reals, because of the way the calculation works.

(4) More sensibly, consider $f: \mathcal{R} \to \mathcal{R}^*$, where $\mathcal{R}^* = \{x \in \mathcal{R} | x > 0\}$, and where $f(z) = e^x$. Now f is a *bijection*, and $g: \mathcal{R}^* \to \mathcal{R}$, with $g(y) = \log y$, is an inverse:

$$e^{\log y} = y,$$

$$\log e^x = x.$$

(5) In this example we assume properties of trigonometric functions. Consider $f: \mathcal{R} \to \mathcal{R}$, $f(x) = \sin x$. This is neither injective nor surjective, so has neither kind of inverse. But what about $\sin^{-1} x$ (or arc sin x) as found in trigonometric tables? The answer depends on exactly what we are trying to achieve. If $\sin^{-1}(x)$ is defined to be the unique y with $-\pi/2 \leq y \leq \pi/2$ such that $\sin y = x$, then this is a right (but not left) inverse to $f: \mathcal{R} \to \{x \in \mathcal{R} | -1 \leq x \leq 1\}$, where $f(x) = \sin x$. It is definitely not a left inverse: for instance

$$\sin^{-1} \sin 6\pi = \sin^{-1} 0 = 0 \neq 6\pi.$$

Sometimes it is said that $\sin^{-1}$ is 'multivalued'. According to our definition, it cannot then be a function, in the legal sense of the term.

(6) The most satisfactory procedure is as follows. Let $f: \{x \in \mathcal{R} \mid -\pi/2 \leqslant x \leqslant \pi/2\} \to \{x \in \mathcal{R} \mid -1 \leqslant x \leqslant 1\}$ be defined by $f(x) = \sin x$. Then f is a bijection, and $\sin^{-1}$ is an inverse function for *this* f.

Left and right inverses need not be unique—this is one reason why their construction involves arbitrary choices. But inverses *are* unique.

PROPOSITION 4.   If a function has both a left inverse and a right inverse, then it has an inverse. This inverse function is unique, and every left or right inverse is equal to it.

PROOF.   If $f: A \to B$ have both a left and a right inverse. Then by proposition 3, f is a bijection, and has an inverse F. If g is any left inverse, then

$$g = g \circ i_B = g \circ (f \circ F) = (g \circ f) \circ F = i_A \circ F = F.$$

Similarly if h is any right inverse then $h = F$. Since an inverse is in particular a left inverse, this also proves F unique.   □

The notation for an inverse function to $f: A \to B$, provided it exists, is

$$f^{-1}: B \to A.$$

PROPOSITION 5.   If $f: A \to B$ and $g: B \to C$ are bijections, then $g \circ f: A \to C$ is a bijection, and

$$(g \circ f)^{-1} = f^{-1} \circ g^{-1}.$$

PROOF.   It is clear that $g \circ f$ is a bijection. It is also easy to verify directly that $f^{-1} \circ g^{-1}$ is a left inverse, since

$$(f^{-1} \circ g^{-1}) \circ (g \circ f) = f^{-1} \circ (g^{-1} \circ (g \circ f))$$
$$= f^{-1} \circ ((g^{-1} \circ g) \circ f)$$
$$= f^{-1} \circ (i_B \circ f)$$
$$= f^{-1} \circ f$$
$$= i_A.$$

Hence by theorem 4 it is an inverse.   □

We can illustrate this as below: it is then clear that the above calculation is much less horrendous than it may appear.

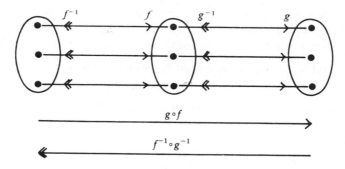

Restriction

If $f: A \to B$ is a function and X is a subset of A we can define a function

$$f|_X: X \to B$$

called the *restriction* of f to X, by

$$f|_X(x) = f(x) \qquad (x \in X).$$

This differs from f only in that we forget about those x which do not lie in X.

For example if $f: \mathcal{R} \to \mathcal{R}$ has $f(x) = \sin x$, and we take $X = \{x \in \mathcal{R} \mid 0 \leqslant x \leqslant 6\pi\}$, then the graphs of f and $f|_X$ are as follows:

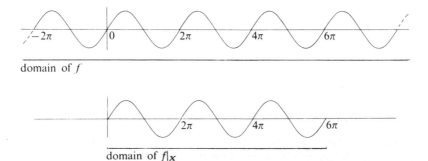

Restriction is a relatively trivial operation to perform. Its main use is to concentrate our attention on how f behaves on the subset X. Sometimes this is useful: we noted above that sin: $\mathscr{R} \to I$ is not a bijection, when $I = \{x \in \mathscr{R} | -1 \leqslant x \leqslant 1\}$. If $X = \{x \in \mathscr{R} | -\pi/2 \leqslant x \leqslant \pi/2\}$ however, then $\sin|_X : X \to I$ is a bijection.

By restricting an identity function $i_A : A \to A$ to a subset $X \subseteq A$ we obtain the *inclusion* function

$$i_A|_X : X \to A$$

for which $i_A|_X(x) = x$ $(x \in X)$. This is therefore the same function as i_X, but with a different choice of codomain which leads to a different emphasis.

Sequences and n-tuples

We can now use functions to tidy up some questions which arose earlier on. In particular we can give precise definitions of sequences and n-tuples. Earlier we have a definition of ordered pairs, triples, quadruples, ... but no general prescription.

Let X_n be the set $\{1, 2, 3, \ldots, n\} = \{x \in \mathscr{N} | 1 \leqslant x \leqslant n\}$. If S is a set then an *n-tuple* of elements of S is defined to be a function

$$f: X_n \to S.$$

What this does is specify $f(1), f(2), \ldots, f(n)$ as elements of S. If we change the notation to $(f_1, f_2, \ldots, f_n)$ we see that two n-tuples $(f_1, \ldots, f_n)$ and $(g_1, \ldots, g_n)$ will be equal if and only if $f_1 = g_1$, $f_2 = g_2, \ldots, f_n = g_n$. This is what an n-tuple should look like.

Similarly the idea of a sequence $a_1, a_2, \ldots$ described earlier as an 'endless list' may be rigorously defined as a function

$$f: \mathscr{N} \to S$$

where now we think of $f(n)$ as a_n.

In the case of ordered pairs the new definition of (f_1, f_2) turns out *not* to be the same as the one given by Kuratowski. However it has the same property, that $(f_1, f_2) = (g_1, g_2)$ if and only if $f_1 = g_1$ and $f_2 = g_2$. Since this is the only property we ever use, the difference is immaterial.

Functions of several variables

In calculus one encounters 'functions of two variables' such as

$$f(x, y) = x^2 - 3y^3 + \cos xy \qquad (x, y \in \mathscr{R}).$$

It is not necessary to go through the whole rigmarole again to make these precise. The very notation makes it clear that f is just an ordinary function defined on the set of ordered pairs (x, y), that is,

$$f: \mathcal{R} \times \mathcal{R} \to \mathcal{R}.$$

In general if A and B are sets then a function of two variables $a \in A$, $b \in B$ is an ordinary function $f: A \times B \to C$. Similarly functions of n variables are ordinary functions defined on a set of n-tuples. These ideas prove to be very fruitful.

Binary operations

A *binary operation* on a set A is simply a function $f: A \times A \to A$. Examples of this concept proliferate in mathematics.

(1) Addition on $\mathcal{N}$, $\alpha: \mathcal{N} \times \mathcal{N} \to \mathcal{N}$, $\alpha(x, y) = x + y$.

(2) Multiplication on $\mathcal{N}$, $\mu: \mathcal{N} \times \mathcal{N} \to \mathcal{N}$, $\mu(x, y) = xy$.

(3) Subtraction on $\mathcal{Z}$, $\sigma: \mathcal{Z} \times \mathcal{Z} \to \mathcal{Z}$, $\sigma(x, y) = x - y$.

(4) Division on the non-zero elements of $\mathcal{Q}$.
Let $\mathcal{Q}^* = \{x \in \mathcal{Q} \mid x \neq 0\}$,
$\delta: \mathcal{Q}^* \times \mathcal{Q}^* \to \mathcal{Q}^*$, $\delta(x, y) = x/y$.

(5) Let $A = \mathbb{P}(X)$, the set of all subsets of a given set X, define
$u: A \times A \to A$ by $u(Y_1, Y_2) = Y_1 \cup Y_2$.

(6) For a given set X, let M be the set of all functions from X to X, so that $f \in M$ means $f: X \to X$. Define $c: M \times M \to M$ by $c(g, f) = g \circ f$.

Most of the occurrences of binary operations in mathematics already have $f(x, y)$ in the form $x \circ y$ for some symbol $\circ$. In the above examples we have $x + y$, xy, $x - y$, x/y, $x \cup y$, $x \circ y$. For this reason we usually denote a binary operation by $\circ: A \times A \to A$ and the image $\circ (x, y)$ by $x \circ y$. After examples (2) and (6), $x \circ y$ is called the *product*, or *composite*, of x and y. In example (2), there is no intervening symbol at all. This economical notation is used often in other mathematical situations where there is no danger of confusion. For instance the composite of functions in (6) is usually written as gf instead of $g \circ f$. The reader has already become accustomed to such conventions when he learned to read

$$2\pi \text{ as } 2 \text{ } times \text{ } \pi,$$

$$2\tfrac{1}{2} \text{ as } 2 \text{ } plus \text{ } \tfrac{1}{2},$$

$$21 \text{ as } 2 \text{ } times \text{ } ten \text{ } plus \text{ } 1.$$

Using the $x \circ y$ notation, we do not expect to find $x \circ y$ and $y \circ x$ to be the same. For example the case of subtraction gives $2-1 \neq 1-2$.

If $x \circ y = y \circ x$ for all $x, y \in A$, then $\circ$ is said to be *commutative*.

Examples (1), (2), and (5) are commutative, (4) and (5) are not. Example (6) is non-commutative if X has more than one element. (If $a, b \in X, a \neq b$, define $f(a) = f(b) = a$, $g(a) = g(b) = b$ and $f(x) = g(x) = x$ otherwise, then $g \circ f(a) = b$, but $f \circ g(a) = a$, so $g \circ f \neq f \circ g$.)

Unless we know that a binary operation is commutative, it is essential to maintain the order of elements in a product. Such a product can be extended to three (and more) elements. Given $x, y, z \in A$, then $x \circ y \in A$ and we may then form the product of this with z. Brackets are introduced at this stage, writing $(x \circ y) \circ z$ to denote the result and to distinguish it from $x \circ (y \circ z)$. Although the latter has x, y, z taken in the same order, it is the product of x and $y \circ z$ and may conceivably be different. For instance $(3-2)-1 \neq 3-(2-1)$.

If $(x \circ y) \circ z = x \circ (y \circ z)$ for all $x, y, z \in A$, then $\circ$ is said to be *associative*. Examples (1), (2), (5), (6) are associative, but (3) and (4) are not.

In developing the number concept, commutative associative binary operations (addition and multiplication) are essential building blocks.

Just as we have binary operations $f: A \times A \to A$, we can go on to define ternary operations $t: A \times A \times A \to A$, and so on. It is even possible to think of a function $g: A \to A$ as a 'unary operation' to begin this hierarchy. Such concepts do not have anything like the central importance of binary operations in mathematics.

Indexed families of sets

At the end of chapter 3 we considered the concept of a set S whose elements were themselves sets. For instance we might have

$$S = \{S_1, \ldots, S_n\}$$

where each S_r is a set. Using the function concept, we can extend this notation. If $\mathcal{N}_n = \{1, 2, \ldots, n\}$, then we have a bijection $f: \mathcal{N}_n \to S$ given by $f(r) = S_r$. There is no reason why we should restrict ourselves to $\mathcal{N}_n$. If A is any set and $f: A \to S$ is a bijection, where every element of S is a set, then we say that S is an *indexed family of sets*,

$$S = \{S_\alpha | \alpha \in A\}.$$

In this situation A is called the *index set*.

The union of such an indexed family,

$$\cup S = \{x \,|\, x \in S_\alpha \text{ for some } \alpha \in A\}$$

is alternatively denoted by

$$\bigcup_{\alpha \in A} S_\alpha$$

and the intersection

$$\cap S = \{x \,|\, x \in S_\alpha \text{ for all } \alpha \in A\}$$

by

$$\bigcap_{\alpha \in A} S_\alpha.$$

When $A = \mathcal{N}_n$, these are often denoted by $\bigcup_{r=1}^{n} S_r$ and $\bigcap_{r=1}^{n} S_r$; when $A = \mathcal{N}$, the notation is $\bigcup_{r=1}^{\infty} S_r, \bigcap_{r=1}^{\infty} S_r$. Don't be worried by the '∞' symbol in these expressions, which are part of the historical development of the subject; the modern notation which omits them is $\bigcup_{r \in \mathcal{N}} S_r$ and $\bigcap_{r \in \mathcal{N}} S_r$.

Exercises

In these exercises, any required properties of exponential, logarithmic, and trigonometric functions may be assumed without proof.

1. Find the images of the following functions $f \colon \mathcal{R} \to \mathcal{R}$.
   (a) $f(x) = x^3$
   (b) $f(x) = x - 4$
   (c) $f(x) = x^2 + 2x + 2$
   (d) $f(x) = x^2 + \cos x$
   (e) $f(x) = 1/x$ if $x \neq 0, f(0) = 1$
   (f) $f(x) = |x|$
   (g) $f(x) = x^2 + x - |x|^2$
   (h) $f(x) = x^{16} + x$.

2. For each of the functions defined above, say whether (as a function $\mathcal{R} \to \mathcal{R}$) it is (a) injective, (b) surjective, (c) bijective.

3. The following functions are to be defined so that their codomain is $\mathcal{R}$, and their domains are certain subsets of $\mathcal{R}$. Say in each case what the largest possible domain is.
   (a) $f(x) = \log x$
   (b) $f(x) = \log \log \cos x$

(c) $f(x) = -x$
(d) $f(x) = \log(1 - x^2)$
(e) $f(x) = \log(\sin^2(x))$
(f) $f(x) = e^{x^2}$
(g) $f(x) = 1/(x^2 - 1)$
(h) $f(x) = \sqrt{[(x - 1)(x - 2)(x - 3)(x - 4)(x - 5)(x - 6)]}$ (positive square root).
Find the image of f in each case.

4. Let S be the set of circles in the plane, and let $f: S \to \mathcal{R}$ be defined by

$$f(S) = \text{the area of } S.$$

Is f injective? Surjective? Bijective?
Now let T be the set of circles in the plane whose centre is the origin, $\mathcal{R}^+ = \{x \in \mathcal{R} | x \geqslant 0\}$, and define $g: T \to \mathcal{R}^+$ by

$$g(T) = \text{the length of the circumference of } T.$$

Is g injective? Surjective? Bijective?

5. If A has two elements and B three, how many different functions are there from A to B? From B to A? How many, in each case, are injective? Surjective? Bijective?

6. If A has n elements and B has m elements ($n, m \in \mathcal{N}$) find the number of functions from A to B.

7. Show that if $A = \varnothing, B \neq \varnothing$, then, according to the set-theoretic definition, there is precisely one function from A to B. Show that if $A \neq \varnothing, B = \varnothing$, there are none. How may functions are there from $\varnothing$ to $\varnothing$?

8. Give examples of functions $f: \mathcal{Z} \to \mathcal{Z}$ which are:
(a) Neither injective nor surjective,
(b) Injective but not surjective,
(c) Surjective but not injective,
(d) Surjective and injective.

9. If $f: A \to B$ show that, for $X \subseteq A$ and $Y \subseteq B$, the formulae

$$\hat{f}(X) = \{f(x) | x \in X\}$$
$$\tilde{f}(Y) = \{x \in A | f(x) \in Y\}$$

define two functions $\hat{f}: \mathbb{P}(A) \to \mathbb{P}(B)$, $\tilde{f}: \mathbb{P}(B) \to \mathbb{P}(A)$.
Prove that for all $X_1, X_2 \subseteq A$ and $Y_1, Y_2 \subseteq B$,
(a) $\hat{f}(X_1 \cup X_2) = \hat{f}(X_1) \cup \hat{f}(X_2)$,
(b) $\hat{f}(X_1 \cap X_2) \subseteq \hat{f}(X_1) \cap \hat{f}(X_2)$ but equality need not hold,
(c) $\tilde{f}(Y_1 \cup Y_2) = \tilde{f}(Y_1) \cup \tilde{f}(Y_2)$,
(d) $\tilde{f}(Y_1 \cap Y_2) = \tilde{f}(Y_1) \cap \tilde{f}(Y_2)$.
Can we improve (b) to equality if further f is known to be surjective? Injective? Bijective?
(In textbooks the usual notation is to write $\hat{f}(X) = f(X)$, $\tilde{f}(Y) = f^{-1}(Y)$: for clarity we have used the notation above.)

10. Define binary operations $\circ$ on $\mathscr{Z}$ by

   (a) $x \circ y = x - y$

   (b) $x \circ y = |x - y|$

   (c) $x \circ y = x + y + xy$

   (d) $x \circ y = \frac{1}{2}(x + y + \frac{1}{2}((-1)^{x+y} + 1) + 1)$.

Verify that these really are binary operations. Which are commutative? Associative?

6

Mathematical logic

THE essential quality of mathematics which binds it together in a coherent way is the use of mathematical proof to deduce new results from known ones, building up a strong and consistent theory. These techniques include some which are unusual in everyday life. Perhaps the most interesting of them is the method of proof by contradiction (or 'reductio ad absurdum' as it was called in more classically oriented times). To show something is true by this method, we assume that it is false and then demonstrate that this assumption leads to a contradiction. For example:

PROPOSITION.   The least upper bound of $S = \{x \in \mathcal{R} | x < 1\}$ is 1.

PROOF.   Certainly 1 is an upper bound. Let K be another upper bound. Suppose $K < 1$; then, by simple arithmetic, $K < \frac{1}{2}(K + 1) < 1$. This means that $K < \frac{1}{2}(K + 1) \in S$, contradicting the fact that K is an upper bound. Thus the assumption that $K < 1$ must be false, so $K \geqslant 1$, and 1 is the least upper bound, as required.   □

This is a typical case of this kind of reasoning. To analyse it more closely, let P stand for the statement 'If K is an upper bound for S, then $K \geqslant 1$'. The major part of the given proof is to establish the truth of P. We assumed P false (i.e. there *is* an upper bound K for S with $K < 1$) and after a simple argument, we derived a contradiction. If the argument is correct, then P cannot be false; so it must be true.

To be able to carry through a proof of this nature and to be certain of its validity, we must make sure of two vital ingredients. In the first place the statement P (and all other statements in the course of the proof, for that matter) must be clearly true or clearly false (though at the time we may not always know which). In everyday conversation we meet comments like 'Almost all drivers exceed the speed limit at some time or other'. This sort of remark would be useless for a contradiction argument. To refute it, is it enough to find just one person who always obeys the speed limit? Do we need to find a 'substantial number' (whatever that may mean precisely) or even a majority? Everyday language is full of such generalities which are vaguely true in most cases, but perhaps not all. Mathematical proof is made of sterner stuff. No such generalities are allowed: all the statements involved must be clearly true or false.

The second essential factor in a proof by contradiction is that the arguments used in the course of a proof must have no flaws. Only if this is so can we be sure in a proof by contradiction that the false link in the chain of argument is the initial assumption: P is false.

There is an old music hall joke which goes something like this:

COMEDIAN:	You're not here.
STRAIGHT MAN:	Don't be silly, of course I am.
COMEDIAN:	You're not, and I'll prove it to you. . . . Look, you're not in Timbuctoo.
STRAIGHT MAN:	No.
COMEDIAN:	You're not at the South Pole.
STRAIGHT MAN:	Of course I'm not.
COMEDIAN:	If you're not in Timbuctoo or at the South Pole, you must be somewhere else.
STRAIGHT MAN:	Of course I'm somewhere else!
COMEDIAN:	Well, if you're somewhere else, you can't be here!

We are amused by this sort of thing and we can all see the logical flaw. But for beginners in mathematical argument it exposes a deep-seated distrust in the use of contradiction arguments. Perhaps something like this happens by accident or sleight of hand in the middle of the proof. When first confronted with a proof that $\sqrt{2}$ is irrational, were you convinced straight away without any degree of suspicion? Such distrust is fully justified: the only way to allay it is to be sure that our mathematical logic is flawless. In the rest of this chapter we will concentrate on the precise use of mathematical language and basic terminology in logic. In the following chapter we return to the techniques of mathematical proof.

Statements

As we have just seen, it is essential that every statement we make in a mathematical proof can be interpreted as being clearly true or false. Typical examples are:

(i) $2 + 3 = 5$.

(ii) The least upper bound of a bounded non-empty subset of $\mathcal{R}$ is unique.

(iii) There is an upper bound K for $S = \{x \in \mathcal{R} \mid x < 1\}$ where $K < 1$.

(iv) $\sqrt{2}$ is irrational.

Here (i), (ii), and (iv) are true, but (iii) is false. In mathematics we are naturally more interested in dealing with true statements than with false ones, but because of the existence of contradiction arguments and suchlike it is convenient for the moment to allow both sides of the coin.

To distinguish between true and false statements we say that each statement has a *truth-value* denoted by the letters t or f, with the obvious interpretation of these symbols: t = true, f = false. Saying that a statement has truth-value t is just a fancy way of saying that it is true.

Given a statement P, the sentence 'P is false' is also a statement which has the opposite truth value to P. For example, if P is the false statement '$2+2=5$', then '$2+2=5$ is false' is a true statement. In logical terminology 'P is false' is usually written

$$\neg P$$

and read as 'not P'. This is a most convenient shorthand notation, although when an actual statement is substituted for P, it may not read grammatically. In the above example, 'not P' would read 'not $2+2=5$'. The equivalent statements '$2+2=5$ is false' or '$2+2 \neq 5$' are more euphonious. When translating 'not P' into words, it is customary to rephrase it in a suitable way to make it read smoothly.

Predicates

A particularly important type of assertion in mathematics is the predicate, introduced in chapter 3. Recall that a predicate is a sentence involving a symbol, such as x, which is either clearly true or clearly false when we replace x by an element of a set X. For instance, a typical mathematical predicate is 'The real number x is not less than 1'. If we denote this by $P(x)$, then $P(2)$ is true, $P(0)$ is false, $P(\pi/4)$ is false, and so on. If we find the truth value of $P(a)$ for every $a \in \mathcal{R}$, we get a *truth function* $T_P : \mathcal{R} \rightarrow \{t, f\}$ for which $T_P(a) = t$ if $P(a)$ is true, and $T_P(a) = f$ if $P(a)$ is false.

This dovetails very nicely with our ideas on set theory. The predicate $P(x)$ partitions $\mathcal{R}$ into two non-overlapping subsets, one containing the elements for which $P(x)$ is true, the other containing the elements for which $P(x)$ is false. The first of these is denoted $\{x \in \mathcal{R} \mid P(x)\}$ in the usual notation. For example $\{x \in \mathcal{R} \mid x \geqslant 1\}$ is the case just described. The other is written $\{x \in \mathcal{R} \mid \neg P(x)\}$ which in the example becomes $\{x \in \mathcal{R} \mid x < 1\}$.

This situation mirrors what happens in general. For any predicate $P(x)$ we get a truth function as above. Then, for $a \in S$, we have

$$a \in \{x \in S | P(x)\} \quad \text{if and only if } P(a) \text{ is true,}$$

$$a \in \{x \in S | \neg P(x)\} \quad \text{if and only if } P(a) \text{ is false.}$$

Rather than use vague remarks like 'a predicate is some sort of statement...' we could use the truth function idea to give a set-theoretic definition. We could start by saying 'a truth function T_P on a set S is any function $T_P: S \to \{t, f\}$. Then we could define 'a predicate $P(x)$ associated with T_P is any sentence equivalent to "$T_P(x) = t$"'. The only trouble with this sort of approach is that seemingly different looking predicates have the same truth function:

$P_1(x)$: 'x is an upper bound for $\{s \in \mathcal{R} | s < 1\}$',

$P_2(x)$: '$x \geqslant 1$'.

It is a major part of a mathematician's job to show that such predicates are equivalent (or, more generally, that the truth of one implies the truth of the other). Thus the predicates dealt with by practising mathematicians do have the structure just described. (In explaining this to the reader, the authors are rather in the position of explaining colour by pointing to something and saying 'that's blue'. The reader is in a better position: he has a fair amount of mathematical experience and should be able to recognize a predicate when he sees one!)

If more than one variable occurs in a sentence, we can talk about a 'predicate in two variables', or 'three variables', and so on. For example the sentence '$x > y$' is a predicate (which we will denote by $Q(x, y)$) in two variables. If real numbers are substituted for x and y then we get a statement. $Q(3, 2)$ is true, $Q(7\frac{1}{4}, 10 + \sqrt{2})$ is false. Here the truth function can be considered as $T_Q: \mathcal{R} \times \mathcal{R} \to \{t, f\}$ where $T_Q(x, y) = t$ if $Q(x, y)$ is true and $T_Q(x, y) = f$ if $Q(x, y)$ is false.

In the same way we can consider '$x^2 + y^2 = z$' as a predicate in three variables $x, y, z \in \mathcal{R}$ which we denote by $R(x, y, z)$. The truth function is $T_R: \mathcal{R} \times \mathcal{R} \times \mathcal{R} \to \{t, f\}$, where

$$T_R(x, y, z) = \begin{cases} t & \text{if} \quad x^2 + y^2 = z \\ f & \text{if} \quad x^2 + y^2 \neq z \end{cases}.$$

In actual practice some mathematicians do not always mention explicitly the set to which a predicate refers, assuming that it is implied by the context. For example the predicate '$x > 3$' is evidently meant to apply to real numbers. It is assumed that no one would dream of substituting something for x which didn't make sense. In the same way it is a time-honoured convention that certain letters invariably stand for elements from a specified set. For example n is usually used to denote a natural number. In this context the predicate '$n > 3$' would be taken to refer only to natural numbers. We have already seen cases of this earlier in the book, for instance in the definition of convergence on page 30 we wrote:

A sequence (a_n) of real numbers tends to a limit l if, given any $\varepsilon > 0$, there is a natural number N such that $|a_n - l| < \varepsilon$ for all $n > N$. Nowhere in this definition do we actually mention that n is a natural number, but it is clearly implied by the context. There is a good reason for doing this, although at first sight it may seem a little sloppy in style. The more explicit one is in mathematics, the more symbols one requires. Taken to ridiculous lengths the page gets so cluttered with symbols that it gets difficult to read the overall meaning because of the mass of detail. It then becomes a question of judgement and mathematical style to select the symbolism which will express the ideas as clearly and succinctly as possible.

All and some

Given a predicate $P(x)$ valid for elements in a set S, we can look to see if it is true for all elements in S, or to see if it is true for at least some elements in S. We can then make the statements 'for all $x \in S$, $P(x)$ is true' or 'for some $x \in S$, $P(x)$ is true'. These can, of course, be true or false statements. They can be written in mathematical symbols using the so-called 'universal quantifier' $\forall$ and the 'existential quantifier' $\exists$.

$\forall x \in S : P(x)$ is read 'for all $x \in S$, $P(x)$'.

$\exists x \in S : P(x)$ is read 'there exists (at least) one $x \in S$ such that $P(x)$'. If the predicate $P(x)$ is true for all $x \in S$, then the statement $\forall x \in S : P(x)$ is true, otherwise it is false. On the other hand, when $P(x)$ is true for at least one $x \in S$, then the statement $\exists x \in S : P(x)$ is true, otherwise it is false. The symbol $\forall x \in S : P(x)$ can be read 'for every $x \in S$, $P(x)$' or 'for each $x \in S$, $P(x)$' or any grammatical equivalent. In the same way $\exists x \in S : P(x)$ can be variously translated as 'there is an $x \in S$ such that $P(x)$', 'for some $x \in S$, $P(x)$' and so on.

It's worth remarking that the statement 'for some $x \in S$, $P(x)$' does not carry with it the connotation that there exist certain other $x \in S$ for which $P(x)$ is false. In ordinary language there are subtle overtones in a statement like:

'some politicians are honest'

as if it implies that there are some who are not. The mathematical usage carries no such implication. Consider the statement:

'some of the numbers 3677, 601, 19, 257, 11119, 7559, 12653 are prime'.

It's easy to see that 19 is prime, so the statement is true. The fact that all the other numbers are also prime does not invalidate this conclusion. At the other end of the scale, 'some' may mean only one; for instance we regard the statement:

'some of the numbers 2, 3, 5, 7, 11 are even'

as being true. This greatly simplifies our task of verifying the truth of $\exists x \in S : P(x)$: we need only find a single value of x for which $P(x)$ is true.

EXAMPLES.   (1) $\forall x \in \mathcal{R} : x^2 \geqslant 0$ means 'for every $x \in \mathcal{R}$, $x^2 \geqslant 0$' or 'the square of any real number is non-negative' or some grammatical equivalent. This is a true statement.

(2) $\exists x \in \mathcal{R} : x^2 \geqslant 0$ reads 'for some $x \in \mathcal{R}$, $x^2 \geqslant 0$' or 'there exists a real number whose square is non-negative'. This is also true.

(3) $\forall x \in \mathcal{R} : x^2 > 0$ is false (since $0^2 \not> 0$).

(4) $\exists x \in \mathcal{R} : x^2 > 0$ is true (since $1^2 > 0$. In this case there are a lot of other elements of $\mathcal{R}$ besides 1 which would do just as well.)

(5) $\exists x \in \mathcal{R} : x^2 < 0$ is false.

If the symbol x is replaced throughout a quantified statement by another symbol, then we regard the new statement as being equivalent to the old.

$\exists x \in S : P(x)$ means the same as $\exists y \in S : P(y)$.

$\forall x \in S : P(x)$ means the same as $\forall y \in S : P(y)$.

For instance $\exists x \in \mathcal{R} : x^2 > 0$ is equivalent to $\exists y \in \mathcal{R} : y^2 > 0$. Both statements say 'there's a real number whose square is positive'.

More than one quantifier

If we have a predicate in two or more variables, then we can use a quantifier for each variable. For example if $P(x, y)$ is the predicate '$x + y = 0$', then the statement $\forall x \in \mathcal{R} \ \exists y \in \mathcal{R} : P(x, y)$ is read 'for every $x \in \mathcal{R}$ there is a $y \in \mathcal{R}$ such that $x + y = 0$'. It is standard logical practice to put all the quantifiers at the front of the predicate and read them in order, for instance $\exists y \in \mathcal{R} \ \forall x \in \mathcal{R} : P(x, y)$ would read 'there is a $y \in \mathcal{R}$ such that for all $x \in \mathcal{R}$, $x + y = 0$'. The order matters. Of the two statements given, $\forall x \in \mathcal{R} \ \exists y \in \mathcal{R} : P(x, y)$ is true (for each $x \in \mathcal{R}$, we can take $y = -x$ to get $x + y = 0$), but $\exists y \in \mathcal{R} \ \forall x \in \mathcal{R} : P(x, y)$ is false, because it asserts the existence of a $y \in \mathcal{R}$ which satisfies $x + y = 0$ for *every* $x \in \mathcal{R}$. No single value of y will do. Getting the order of the quantifiers right in such a statement is a vital part of clear mathematical thinking. It is a common error to get it wrong (and not just among beginners either!). A source of this problem is when we come to write what is probably a clear logical statement in flowing mathematical prose. The word order is changed around somewhat to give a more euphonious sound to the language, sometimes at the expense of logical clarity. In particular the quantifiers become embedded in the middle of the sentence instead of all coming at the beginning. (We have already done this a few lines above when we wrote '. . . it asserts *the existence* of a $y \in \mathcal{R}$ which satisfies $x + y = 0$ *for every* $x \in \mathcal{R}$!') Consider the following:

'Every non-zero rational number has a rational inverse'.

By this statement we mean that 'Given $x \in \mathcal{Q}$, where $x \neq 0$, there is an element $y \in \mathcal{Q}$ such that $xy = 1$.' This is of course true: if $x = p/q$ where p, q are integers with $p \neq 0$, then we can take $y = q/p$. Written in logical language it becomes:

$$\text{'}\forall x \in \mathcal{Q} \ (x \neq 0) \ \exists y \in \mathcal{Q} : xy = 1\text{'}.$$

A mathematician might change the order and say 'There's a rational inverse for every non-zero rational number' to convey the same idea, even though this kind of statement could be misinterpreted. You can help matters by making sure that the meaning of your written mathematics is as clear as you can possibly make it.

The ambiguity only arises when the quantifiers involved are different. If they are the same, there is no such problem. For instance, given the predicate $P(x, y) : \text{'}(x + y)^2 = x^2 + 2xy + y^2\text{'}$, then the two statements

$$\forall x \in \mathcal{R} \ \forall y \in \mathcal{R} : P(x, y)$$

and

$$\forall y \in \mathcal{R} \ \forall x \in \mathcal{R} : P(x, y)$$

both amount to the same thing 'for all $x, y \in \mathcal{R}$, $(x+y)^2 = x^2 + 2xy + y^2$', which is of course a true statement. If the variables involved come from the same set, as in this case, we usually simplify the notation, writing $\forall x, y \in \mathcal{R}: P(x, y)$. The same happens with the existential quantifier. For instance if $Q(x, y)$ is 'x, y are irrational and $x + y$ is rational', then $\exists x \in \mathcal{R} \backslash \mathcal{Q} \; \exists y \in \mathcal{R} \backslash \mathcal{Q}: Q(x, y)$ and $\exists y \in \mathcal{R} \backslash \mathcal{Q} \; \exists x \in \mathcal{R} \backslash \mathcal{Q}: Q(x, y)$ both say 'there exist two real numbers x, y which are irrational but whose sum is rational'. (The latter is a true statement since $\sqrt{2}$ and $-\sqrt{2}$ are irrational, but 0 is rational.) This may also be written as $\exists x, y \in \mathcal{R} \backslash \mathcal{Q}: Q(x, y)$.

There is another minor pitfall in written mathematics. That is that the universal quantifier is not always explicitly written; often it is implied by the context. Take another look at the definition of convergence of a sequence on page 30:

A sequence of real numbers tends to a limit l if, given any $\varepsilon > 0$, there is a natural number N such that $|a_n - l| < \varepsilon$ for all $n > N$.

This is quite a mouthful to say and various mathematicians try to cut down the verbiage in some way or other to make it as brief as possible. One of the little words that gets lost is 'all'. A typical statement of the definition of $a_n \to l$ is:

Given $\varepsilon > 0$, $\exists N$ such that $n > N$ implies $|a_n - l| < \varepsilon$.

You will find a lot of minor variations on this definition, but in essence they all mean the same thing. If you understand this, you are a long way along the road to understanding the nature of the problem of communicating mathematics with the appropriate degree of precision.

The negation of a statement

On page 111 we introduced the negation $\neg P$ of a statement P.

The truth-value of $\neg P$ can be represented in the following table (called a *truth table*):

P	$\neg P$
t	f
f	t

Reading along the rows, this simply says that when P is true, $\neg P$ is false and vice versa. The symbol $\neg$ is called a *modifier* because it modifies a statement, changing its meaning and its truth value.

In the same way a predicate can be modified using $\neg$. If $P(x)$ is '$x > 5$', then $\neg P(x)$ is '$x > 5$ is false' or equivalently, '$x \not> 5$'.

The negation of a statement involving quantifiers leads us to an interesting situation. It is easy to see that the statement '$\forall x \in S : P(x)$ is false' is the same thing as '$\exists x \in S : \neg P(x)$'. (If it is false that $P(x)$ is true for all $x \in S$, then there must exist an $x \in S$ for which $P(x)$ is false, in which case $\neg P(x)$ is true.) This can be written symbolically as:

(1) $\neg \forall x \in S : P(x)$ means the same as $\exists x \in S : \neg P(x)$.

Similarly

(2)   . $\neg \exists x \in S : P(x)$ means the same as $\forall x \in S : \neg P(x)$.

The second of these states that to say 'there is no x for which $P(x)$ is true' is the same as saying 'for every $x \in S$, $P(x)$ is false'. As an example of (2) we have:

$\neg \exists x \in \mathcal{R} : x^2 < 0$ ... there is no $x \in \mathcal{R}$ such that $x^2 < 0$.

$\forall x \in \mathcal{R} : \neg (x^2 < 0)$ ... every $x \in \mathcal{R}$ satisfies $x^2 \not< 0$.

These two simple principles are of vital importance in mathematical argument. Freely translated, the first says 'to show that a predicate $P(x)$ is not true for all $x \in S$, it is only necessary to exhibit one x for which $P(x)$ is false'. The second asserts 'to show no $x \in S$ exists for which $P(x)$ is true, it is necessary to prove $P(x)$ false for every $x \in S$'.

As rules of thumb for negating statements involving quantifiers, they really come into their own when several quantifiers are used. A typical instance is the definition of convergence of a sequence; in logical notation this is:

$$\forall \varepsilon > 0 \; \exists N \in \mathcal{N} \; \forall n > N(|a_n - l| < \varepsilon).$$

To show that a sequence (a_n) does *not* tend to a limit l, we have to prove the negation of this statement, i.e.

$$\neg [\forall \varepsilon > 0 \; \exists N \in \mathcal{N} \; \forall n > N(|a_n - l| < \varepsilon)].$$

Using the principles (1) and (2) this becomes successively

$$\exists \varepsilon > 0 \neg [\exists N \in \mathcal{N} \; \forall n > N(|a_n - l| < \varepsilon)],$$

then

$$\exists \varepsilon > 0 \; \forall N \in \mathcal{N} \neg [\forall n > N (|a_n - l| < \varepsilon)],$$

then

$$\exists \varepsilon > 0 \forall N \in \mathcal{N} \exists n > N \neg (|a_n - l| < \varepsilon),$$

which translates finally into:

$$\exists \varepsilon > 0 \forall N \in \mathcal{N} \exists n > N (|a_n - l| \nless \varepsilon).$$

To verify that (a_n) does not converge to l, we have to prove this last line true, i.e. we have to show that there is a particular $\varepsilon > 0$ such that for any natural number N there is always a larger natural number $n > N$ with $|a_n - l| \nless \varepsilon$. Much of the difficulty in a subject like mathematical analysis is in manipulating statements like this. It becomes much easier with a little experience and practice, always keeping the principles of negating quantifiers in mind.

Logical grammar: connectives

In mathematics we use standard conjunctions like 'and', 'or' etc. with very specific meanings. For instance 'or' is used in the 'inclusive' sense. This means that if P, Q are specific statements, then P or Q is a statement which is regarded as true provided that one or both of P, Q is true. We can represent this by a truth-table:

P	Q	P or Q
t	t	t
t	f	t
f	t	t
f	f	f

This is read along the horizontal rows. For example, the second row says that if P is true, Q is false, then P or Q is true.

Other conjunctions in regular use in mathematics are 'and', 'implies', and 'if and only if'. These are given the appropriate symbols: & (and), $\Rightarrow$ (implies), $\Leftrightarrow$ (if and only if). They have the following truth tables:

P	Q	P & Q
t	t	t
t	f	f
f	t	f
f	f	f

P	Q	P ⇒ Q
t	t	t
t	f	f
f	t	t
f	f	t

P	Q	P ⇔ Q
t	t	t
t	f	f
f	t	f
f	f	t

These tables are read in the same way. The first and last of these are fairly obvious. $P \& Q$ is only regarded as true when *both* P and Q are true, $P \Leftrightarrow Q$ is regarded as true only when P and Q each have the same truth value. The interesting table is the one for $P \Rightarrow Q$. If P is true, then the first and second lines say that the implication $P \Rightarrow Q$ is true when Q is true and false when Q is false. This shows that the truth of $P \Rightarrow Q$ means that if P is true, then Q must be true. This is the normal interpretation of the implication sign $\Rightarrow$, and for this reason $P \Rightarrow Q$ is often interpreted as 'if P, then Q'. But what of the situation when P is false? The third and fourth lines say that whether Q is true or false, then $P \Rightarrow Q$ is regarded as being true. In many places a lot of philosophical nonsense is talked about this. 'How can the falsehood of P imply the truth of Q?' The reason for this situation can be seen in the standard mathematical practice of using connectives with predicates rather than statements. If $P(x)$ and $Q(x)$ are predicates both valid for $x \in S$, then we can use the connectives in the manner above to get predicates $P(x)$ or $Q(x)$, $P(x) \& Q(x)$ etc. In particular we have the predicate $P(x) \Rightarrow Q(x)$ with the given truth table. Sometimes the predicate $P(x) \Rightarrow Q(x)$ is true for all $x \in S$. This is where the truth table comes into its own. For example when $P(x)$ is '$x > 5$' and $Q(x)$ is '$x > 2$', then every mathematician would agree that $P(x) \Rightarrow Q(x)$ is true, although some would read this as '*if* $x > 5$, *then* $x > 2$', they would not be interested in what happens when $x \not> 5$. Let us substitute some different values for x and see what happens:

If $x = 4$, then $P(4)$ is false and $Q(4)$ is true.

If $x = 1$, then $P(1)$ is false and $Q(1)$ is false.

These are precisely lines three and four of the truth table for '$\Rightarrow$', and illustrate how the truth table is arrived at. With this interpretation, the truth table can best be described as follows:

$$\text{`} P \Rightarrow Q \text{ is true'}$$

means that

(a) 'If P is true, then Q must be true',

however,

(b) 'If P is false then Q may be either true or false, and no conclusion may be drawn in this case'.

Other connectives are possible, for example the 'exclusive or' (denoted here by OR) with truth-table:

P	Q	P OR Q
t	t	f
t	f	t
f	t	t
f	f	f

P OR Q is true when one, but not both, of P, Q is true.

Technically one could write down truth tables for other connectives, but these can be manufactured by combining the given ones. For instance the exclusive OR involved in the statement P OR Q can also be described using the statement $(P$ or $Q)$ & $\neg(P$ & $Q)$. This will be discussed in greater detail in the section on 'formulae for compound statements'.

In writing mathematics, the practising mathematician does not restrict himself stylistically to using the connectives just described. He will use grammatical connectives like 'but', 'since', 'because' etc. as his fancy takes him. These are all to be interpreted in the obvious way as grammatical equivalents for the given words. For instance the truth table for 'P but Q' is the same as that for 'P & Q'. The statement '$\sqrt{2}$ is irrational but $(\sqrt{2})^2$ is rational' means the same thing as '$\sqrt{2}$ is irrational and $(\sqrt{2})^2$ is rational'. Similarly 'P because Q' and 'P since Q' have the same truth table as '$Q \Rightarrow P$'. The reader can make himself familiar with these variants by looking at a few examples. (See the exercises at the end of the chapter.)

The link with set theory

If we apply the connectives and modifier $\neg$ to predicates in one variable, then we get a simple relationship with set-theoretic notation. Suppose $P(x)$ and $Q(x)$ are predicates valid on the same set S and we look at the subsets for which the various compound statements are true. For '&' we obtain:

$$\{x \in S | P(x) \text{ \& } Q(x)\} = \{x \in S | P(x)\} \cap \{x \in S | Q(x)\}.$$

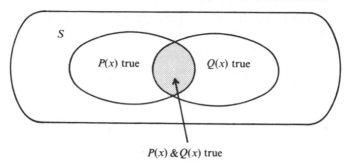

$P(x) \& Q(x)$ true

Similarly $\{x \in S | P(x) \text{ or } Q(x)\} = \{x \in S | P(x)\} \cup \{x \in S | Q(x)\}$.

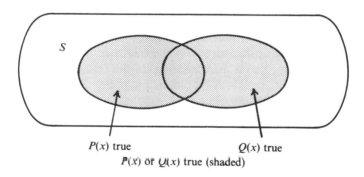

$P(x)$ true $Q(x)$ true
$P(x)$ or $Q(x)$ true (shaded)

This is one reason for using the 'inclusive or', corresponding to set-theoretic union, rather than the 'exclusive OR' which corresponds to the so-called 'symmetric difference' in set theory, represented by the shaded area in the diagram:

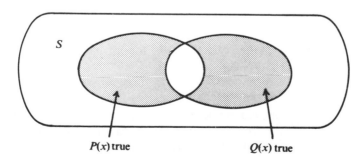

$P(x)$ true $Q(x)$ true

The modifier $\neg$ applied to a single predicate $P(x)$ corresponds to the set-theoretic complement:

$$\{x \in S | \neg P(x)\} = S\backslash\{x \in S | P(x)\}.$$

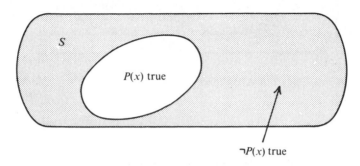

When we come to consider the implication $P(x) \Rightarrow Q(x)$ we look at the situation a little differently. We are really only interested in the case where $P(x) \Rightarrow Q(x)$ is true for all x. In this case if $P(x)$ is true, so must $Q(x)$ be, i.e. if $a \in \{x \in S | P(x)\}$ then $a \in \{x \in S | Q(x)\}$ which means $\{x \in S | P(x)\} \subseteq \{x \in S | Q(x)\}$. The truth of the statement $P(x) \Rightarrow Q(x)$ corresponds to set-theoretic inclusion.

$$P(x) \Rightarrow Q(x) \text{ true for all } x \in S:$$

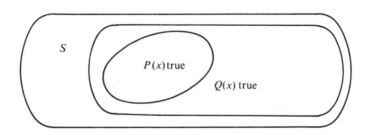

In the same way $P(x) \Leftrightarrow Q(x)$ is true for all $x \in S$ if and only if $\{x \in S | P(x)\} = \{x \in S | Q(x)\}$.

Formulae for compound statements

Using connectives and modifiers we can form more complex statements and predicates from given ones, for instance $(P \ \& \ Q)$ or R. This involves three statements and the truth table has $2^3 = 8$ lines:

P	Q	R	Intermediate calculation P & Q	(P & Q) or R
t	t	t	t	t
t	t	f	t	t
t	f	t	f	t
t	f	f	f	f
f	t	t	f	t
f	t	f	f	f
f	f	t	f	t
f	f	f	f	f

Notice that the symbol '$(P \,\&\, Q)$ or R' is really a recipe for forming a new statement or predicate out of three given ones. To emphasize this we will call it a *compound statement formula* when we think of P, Q, R, as standing for unspecified statements or predicates. When we replace P, Q, R by specific statements, for instance

$$(2 > 3 \,\&\, 2 > 6) \text{ or } 2 > 1,$$

we will call it a *compound statement*. If instead we use specific predicates, we shall call the result a *compound predicate*. For example,

$$(x > 3 \,\&\, x > 6) \text{ or } x > 1$$

is a compound predicate.

A large part of any mathematical proof is involved with manipulating compound statements and predicates. In building them up, brackets are often essential to show how they are constructed. For instance $P \,\&\, (Q \text{ or } R)$ is different from $(P \,\&\, Q)$ or R. In fact, looking at the seventh line in the above truth table: if P is false, Q is false, and R is true, then $(P \,\&\, Q)$ or R is true; but a calculation shows that in this case $P \,\&\, (Q \text{ or } R)$ is false. The same goes for predicates. So we must take care to put the brackets in the right places whenever ambiguities would arise. (Sometimes it is permissible to omit them. For instance, $(P \,\&\, Q) \,\&\, R$ has the same truth table as $P \,\&\, (Q \,\&\, R)$, so it would not cause problems to write just $P \,\&\, Q \,\&\, R$.)

When we build up compound statement formulae using connectives and modifiers we often find formulae which look different, but have the same truth table. For instance, $P \Rightarrow Q$ and $(\neg Q) \Rightarrow (\neg P)$:

P	Q	$P \Rightarrow Q$
t	t	t
t	f	f
f	t	t
f	f	t

P	Q	$\neg P$	Intermediate calculations $\neg Q$	$(\neg Q) \Rightarrow (\neg P)$
t	t	f	f	t
t	f	f	t	f
f	t	t	f	t
f	f	t	t	t

or, omitting the intermediate calculations,

P	Q	$(\neg Q) \Rightarrow (\neg P)$
t	t	t
t	f	f
f	t	t
f	f	t

In this case the compound statement formulae are said to be *logically equivalent*.

Denoting two compound statement formulae by S_1, S_2, we write

$$S_1 \equiv S_2$$

for logical equivalence. For instance, our result above can be expressed as

$$P \Rightarrow Q \equiv (\neg Q) \Rightarrow (\neg P).$$

An interesting fact about compound statement formulae is that sometimes two of them can be considered equivalent even though they are composed of different symbols. This happens when changing the truth value of a particular symbol does not affect the final result. For instance $P \& (\neg P)$ is always false. Working out the truth table of $(P \& (\neg P))$ or $(\neg Q)$ we have

P	Q	$(P \& (\neg P))$ or $(\neg Q)$
t	t	f
t	f	t
f	t	f
f	f	t

so $(P \& \neg P))$ or $(\neg Q)$ always has the same truth value as $\neg Q$, regardless of the truth value of P. One way of looking at this, typical of the mathematical fraternity, is to think of $\neg Q$ as a function of both P and Q, so that its truth table becomes

P	Q	$\neg Q$
t	t	f
t	f	t
f	t	f
f	f	t

By this formal device we can legitimately write

$$\neg Q \equiv (P \& (\neg P)) \text{ or } (\neg Q).$$

A compound statement formula is called a *tautology* if it is true regardless of the truth values of its constituent statement symbols. Typical examples are:

(i) P or $(\neg P)$

(ii) $P \Rightarrow (P \text{ or } Q)$

(iii) $(P \& Q) \Rightarrow P$

(iv) $(P \Rightarrow Q) \Leftrightarrow ((\neg Q) \Rightarrow (\neg P))$.

The reader should check that the truth tables for these always yield the value t.

If a compound statement formula always takes truth value f, regardless of the truth values of its constituent statement symbols, then it is called a *contradiction*. For example $P \& (\neg P)$ is a contradiction.

Any two tautologies are logically equivalent, and any two contradictions are logically equivalent. Moreover, a compound statement formula S is a tautology if and only if $\neg S$ is a contradiction.

It is useful to use the symbol T for a tautology and C for a contradiction. We then get some interesting results. For example, $(\neg P) \Rightarrow C$ is logically equivalent to P. The truth tables are

P	$(\neg P) \Rightarrow C$		P	P
t	t		t	t
f	f		f	f

(In calculating the first table, remember that C always has truth value f.)

If we replace C by a specific contradiction, say $Q \mathbin{\&} (\neg Q)$, we will still get the same result:

P	Q	$(\neg P) \Rightarrow (Q \mathbin{\&} (\neg Q))$		P	Q	P
t	t	t		t	t	t
t	f	t		t	f	t
f	t	f		f	t	f
f	f	f		f	f	f

The reader should check all the intermediate calculations in the first table to get a feeling for what is happening.

Instead of comparing the truth tables of two compound statement formula S_1 and S_2 to check logical equivalence, we can look at the single table for $S_1 \Leftrightarrow S_2$. If S_1 is logically equivalent to S_2, then $S_1 \Leftrightarrow S_2$ is a tautology, and vice versa. For example, the logical equivalence of $P \Rightarrow Q$ and $(\neg Q) \Rightarrow (\neg P)$ corresponds to the fact that $[P \Rightarrow Q] \Leftrightarrow [(\neg Q) \Rightarrow (\neg P)]$ is a tautology.

Logical deductions

The overall strategy behind a proof often arises by proving, not the truth of a given statement, but the truth of a logically equivalent one. Important examples are as follows.

(1) *The contrapositive* $P \Rightarrow Q \equiv (\neg Q) \Rightarrow (\neg P)$. To prove $P \Rightarrow Q$, we establish the truth of $(\neg Q) \Rightarrow (\neg P)$.

(2) *Proof by contradiction* $P \equiv (\neg P) \Rightarrow C$ where C is a contradiction. To prove P, we establish the truth of $(\neg P) \Rightarrow C$.

(3) *'If and only if' proof* $P \Leftrightarrow Q \equiv (P \Rightarrow Q) \mathbin{\&} (Q \Rightarrow P)$.

(4) *'If and only if' version II* $P \Leftrightarrow Q \equiv (P \Rightarrow Q) \mathbin{\&} ((\neg P) \Rightarrow (\neg Q))$. To prove $P \Leftrightarrow Q$, we establish the truth of both $P \Rightarrow Q$ and $(\neg P) \Rightarrow (\neg Q)$.

We establish the truth of a given statement from known ones by seeing how the new statement is made up, and using truth tables. For instance, we might know that P is true and that $(\neg Q) \Rightarrow (\neg P)$ is true. From these facts we can deduce that Q must be true. The given statements might be compound ones, like $(\neg Q) \Rightarrow (\neg P)$, and although we know that the total statement is true, we may have no information on the truth of its constituents. Thus we might know that $(\neg Q) \Rightarrow (\neg P)$ is true, but have no knowledge of the truth values of P or of Q. This still allows us to make some deductions; for example if $(\neg Q) \Rightarrow (\neg P)$ is true, then we know that the equivalent statement $P \Rightarrow Q$ is true. Here are a few situations in which we can deduce the truth of the statement in the second column from those in the first.

If these statements are true . . .	*. . . then this must be true*
$P, (\neg Q) \Rightarrow (\neg P)$	Q
$(\neg P) \Rightarrow C$ (contradiction)	P
$P, P \Rightarrow Q$	Q
$P \Rightarrow Q, Q \Rightarrow R$	$P \Rightarrow R$
P or $Q, \neg P$	Q
$P \& Q$	P or Q
$P \Rightarrow Q, Q \Rightarrow P$	$P \Leftrightarrow Q$
$P_1, \ldots, P_n$	$P_1 \& \ldots \& P_n$
$P_1, \ldots, P_n, (P_1 \& \ldots \& P_n) \Rightarrow Q$	Q

This table can be continued ad infinitum. To obtain a new entry, write a number of compound statement formulas $S_1, \ldots, S_n$ in the left hand column. In the corresponding position in the right hand column put any compound statement formula D whose truth is ensured when $S_1, \ldots, S_n$ are true. This involves looking at the truth tables for $S_1, \ldots, S_n, D$; but we can formulate the condition in one composite table by considering the formula $(S_1 \& \ldots \& S_n) \Rightarrow D$. If this is a *tautology* then the truth of $S_1, \ldots, S_n$ ensures the truth of D, as required.

A tautology of the form $(S_1 \& \ldots \& S_n) \Rightarrow D$ is called a *rule of inference*. Given such a rule of inference, and substituting actual statements into the compound statement formulae involved, if $S_1, \ldots, S_n$ are true, then we may infer that D is true.

When the statements concerned involve quantified predicates we must look at how they are composed to see if the truth of one statement is a natural consequence of given ones. In a simple case we might know that $\forall x \in S : P(x)$ is true, and infer from this that $\exists x \in S : P(x)$ is also true. Given the truth of $\forall x \in S : P(x)$ and $\forall x \in S : Q(x)$ we can deduce a whole host of statements, including $\forall x \in S: P(x) \,\&\, Q(x)$, $[\forall x \in S: P(x)]$ or $[\forall x \in S: Q(x)]$, $P(a) \,\&\, Q(b)$ where $a, b \in S$, and so on. Again, we can make a list of deductions which can be made from statements involving quantified predicates.

If these statements are true . . .	*. . . then this must be true*
$\forall x \in S: P(x),\ \forall x \in S: Q(x)$	$\forall x \in S: [P(x) \,\&\, Q(x)]$
$\forall x \in S: P(x)$	$\exists x \in S: P(x)$
$\forall x \in S: P(x)$	$P(a)\ (a \in S)$
$P(a)\ (a \in S)$	$\exists x \in S: P(x)$
$\neg[\forall x \in S: P(x)]$	$\exists x \in S: [\neg P(x)]$
$\neg[\exists x \in S: P(x)]$	$\forall x \in S: [\neg P(x)]$
$\exists x \in S\ \forall y \in S\ \forall z \in M: \neg[P(x, y, z)]$	$\neg[\forall x \in S\ \exists y \in S\ \exists z \in M: P(x, y, z)]$
etc.	etc.

This list could be continued with a large variety of other deductions. In the left-hand column we put statements $S_1, \ldots, S_n$ which may involve quantified predicates and in the right-hand column we put a statement D whose truth follows when all of $S_1, \ldots, S_n$ are true. Again this can be formulated as the requirement that the single statement $(S_1 \,\&\, \ldots \,\&\, S_2) \Rightarrow D$ must be true. Predicates are much more complex things than plain statements, and we shall not formulate a *general* test for this.

Proof

In practice we start from a number of statements $H_1, \ldots, H_r$ and wish to deduce the truth of a statement D. The process may become quite involved, with the introduction of other subsidiary statements. For this reason we perform the process in a number of steps by writing down a sequence of statements $L_1, \ldots, L_n$, where $L_n = D$, and each line L_m is either one of the hypotheses $H_1, \ldots, H_r$ or its truth can be deduced from the truth of $L_1, \ldots, L_{m-1}$, for each $m = 1, 2, \ldots, n$. In particular L_1 must be one of the hypotheses and each successive line $L_2, \ldots, L_n$ must either be a true deduction from previous lines, or a hypothesis. This clearly implies the truth of the last line D. The truth of the deduction of L_m from previous lines is checked, as before, by verifying the truth of $(L_1 \,\&\, \ldots \,\&\, L_{m-1}) \Rightarrow L_m$. In the case that L_m is a hypothesis it follows immediately from the truth table for $\Rightarrow$ that $(L_1 \,\&\, \ldots \,\&\, L_{m-1}) \Rightarrow L_m$ is true; but when L_m is not a hypothesis we need to check more fully.

When the final deduction D is of the form $P \Rightarrow Q$, then mathematicians often vary the prescription by writing down lines $L_1, \ldots, L_n$ with

P as the first line L_1 and Q as the last line L_n. Here each intermediate line must either be a hypothesis, or its truth must follow from previous lines, as before. Some lines may well be predicates: again the important thing is to ensure that $(L_1 \& \ldots \& L_{m-1}) \Rightarrow L_m$ is always true.

EXAMPLE.   Given hypotheses
H_1: $5 > 2$,
H_2: $\forall x, y, z \in \mathcal{R} : (x > y) \& (y > z) \Rightarrow (x > z)$,
we can write down the proof that $(x > 5) \Rightarrow (x > 2)$ as:
L_1: $x > 5$
L_2: $5 > 2$
L_3: $\forall x, y, z \in \mathcal{R} : (x > y) \& (y > z) \Rightarrow (x > z)$
L_4: $x > 2$.
Although this particular deduction is not exactly mindboggling, it embodies the general prescription for a proof which we crystallize as follows:

DEFINITION.   A *proof* of the statement $P \Rightarrow Q$ (where P and Q may be statements or predicates), given the statements $H_1, \ldots, H_r$, consists of a finite number of lines

$$L_1 = P$$

$$L_2$$

$$\vdots$$

$$L_n = Q$$

where each line L_m $(2 \leqslant m \leqslant n)$ is either a hypothesis H_s $(1 \leqslant s \leqslant r)$ or a statement or predicate such that

$$(L_1 \& \ldots \& L_{m-1}) \Rightarrow L_m$$

is a true statement.

Under these conditions, if P is true, then each succeeding line must also be true, so in particular the truth of Q will follow. From the truth table for $\Rightarrow$, we see that this establishes the truth of $P \Rightarrow Q$.

It is worth looking at what happens when P is false. This could easily happen if P is a predicate, when substituting particular values for the variable renders the predicate untrue. Thus in the above example, $x > 5$ is untrue when $x = 1$ (for example), in which case the line L_4 becomes $1 > 2$ which is also false. On the other hand if x were 3 then L_1 would be false but L_4 would be $3 > 2$ which is true. In short, if P is false we can draw no conclusions as to the validity or otherwise of succeeding lines: the only thing we are certain of is that the *compound* statement $P \Rightarrow Q$ is true. This is because, although we know that the deductions $(L_1 \& \ldots \& L_{m-1}) \Rightarrow L_m$ are all true, the falsity of L_1 can lead to the falsity of L_m.

This is the most important factor in proof by contradiction. It has exactly the same format as the one given above. To establish P we prove an equivalent statement $(\neg P) \Rightarrow C$, where C is some contradiction. So we start with the first line L_1 as $(\neg P)$ and end up with the last line L_n being C. On the assumption $(\neg P)$ is true, each succeeding line must also be true. But L_n is manifestly false, being a contradiction. Hence $(\neg P)$ cannot be true, so P must be true. In this way we establish the truth of P 'by contradiction'.

The proof of $P \Rightarrow Q$ by establishing the logically equivalent statement $(\neg Q) \Rightarrow (\neg P)$ also has the same basic structure, starting with the line $(\neg Q)$, and ending with the line $(\neg P)$.

This is the formal definition of the logical steps in a proof. What do we actually need to write down in practice? The next chapter provides a possible answer.

Exercises

1. Write down the truth-tables for the following compound statements:
    (a) $P \Rightarrow (\neg P)$
    (b) $((P \Rightarrow R) \& (Q \Rightarrow R)) \Leftrightarrow ((P \& Q) \Rightarrow R)$
    (c) $(P \& Q) \Rightarrow (P \text{ or } Q)$
    (d) $(P \Rightarrow Q) \text{ or } (Q \Rightarrow P) \text{ or } (\neg Q)$.
Which are tautologies?

2. Write the following statements using the quantifiers $\forall$, $\exists$, and state which of them are true:
    (a) For every real number x there exists a real number y such that $y^3 = x$.
    (b) There exists a real number y such that for every real number x, the sum $x + y$ is positive.
    (c) For each irrational number x, there is an integer n satisfying $x < n < x + 1$.
    (d) The square of every integer leaves remainder 0 or 1 on division by 4.
    (e) The sum of the squares of two prime numbers which are not equal to 2 is an even number.

3. Translate the following statements into prose:
    (a) $\forall x \in \mathscr{R} \ \exists y \in \mathscr{R} : x^2 - 3xy + 2y^2 = 0$.
    (b) $\exists y \in \mathscr{R} \ \forall x \in \mathscr{R} : x^2 - 3xy + 2y^2 = 0$.
    (c) $\exists N \in \mathscr{N} \forall \varepsilon \in \mathscr{R} : [(\varepsilon > 0) \& (n > N)] \Rightarrow (1/n < \varepsilon)$.
    (d) $\forall x \in \mathscr{N} \ \forall y \in \mathscr{N} \ \exists z \in \mathscr{N} : x + z = y$.
    (e) $\forall x \in \mathscr{Z} \ \forall y \in \mathscr{Z} \ \exists z \in \mathscr{Z} : x + z = y$.

Read your translations carefully, and if you think they sound stilted, rewrite them in a more flowing style (but don't change their meaning!). State which of (a)–(e) are true and which are false, giving a reason in each case.

4. In each of the following cases, write out truth tables and say whether the two statements are equivalent or not.

(a) $\neg[P \& (\neg P)]$, P or $(\neg P)$

(b) $P \Rightarrow Q$, $(\neg P) \& Q$

(c) $P \Rightarrow Q$, $(\neg Q) \& P$

(d) $(P \Rightarrow Q) \& R$, $P \Rightarrow (Q \& R)$

(e) $[P \& (\neg Q)] \Rightarrow [R \& (\neg R)]$, $P \Rightarrow Q$.

5. Use truth tables (where possible) to verify the rules of inference listed in the section 'logical deductions'.

6. Which of the following are logically correct deductions?

(a) 'If the Strategic Arms Limitation Agreement is signed, or the United Nations approve a disarmament plan, then shares in the arms industry will slump. But armament shares will not slump, so the Strategic Arms Limitation Agreement will not be signed.'

(b) 'If Britain leaves the EEC or if the trade deficit is reduced, the price of butter will fall. If Britain stays in the EEC exports will not increase. The trade deficit will increase unless exports are increased. Therefore the price of butter will not fall.'

(c) 'Some politicians are honest. Some women are politicians. Therefore some women politicians are honest.'

(d) 'If I do not work hard I will sleep. If I am worried I will not sleep. Therefore if I am worried I will work hard.'

7. Consider the statement

$$x \leqslant y \quad \text{but} \quad y > z$$

in the following cases:

(a) $x = 1, y = 2, z = 0$

(b) $x = 1, y = 2, z = 3$

(c) $x = 2, y = 1, z = 0$

(d) $x = 2, y = 1, z = 3$.

Which cases yield a true statement? Use this information to draw up a truth table for 'but' and check that

$$(P \text{ but } Q) \Leftrightarrow (P \& Q)$$

is a tautology.

Do the same for 'since' and 'therefore' and compare with 'implies'. What happens for 'unless'?

8. What are the negations of the following statements?

(a) $\forall x : (P(x) \& Q(x))$

(b) $\exists x : (P(x) \Rightarrow Q(x))$

(c) $\forall x \in \mathcal{R} \; \exists y \in \mathcal{R} : x \geqslant y$

(d) $\forall x \in \mathcal{R} \; \forall y \in \mathcal{R} \; \exists z \in \mathcal{Q} : x + y \geqslant z$.

Are (c) and (d) true or false?

9. Prove by contradiction arguments the following theorems.

(a) If $x, y \in \mathcal{R}$ and $y \leqslant x + \varepsilon$ for all $\varepsilon > 0$ ($\varepsilon \in \mathcal{R}$), then $y \leqslant x$.

(b) For all real numbers x, either $\sqrt{3} + x$ is irrational or $\sqrt{3} - x$ is irrational.

(c) There is no smallest rational number greater than $\sqrt{2}$.

10. Consider the connectives $\neg$, &, or $\Rightarrow$. Show that

$$P \Rightarrow Q \equiv (\neg P) \,\&\, Q$$

$$P \text{ or } Q \equiv \neg[(\neg P) \,\&\, (\neg Q)]$$

and hence that any compound statement can be written in terms of the connectives $\neg$, & alone. Is it possible to write every compound statement in terms of just one of the connectives $\neg$, &, or, $\Rightarrow$?

Define the *stroke* connective $|$ by the truth table

P	Q	$P|Q$
t	t	f
t	f	t
f	t	t
f	f	t

and show that

$$P|Q \equiv (\neg P) \text{ or } (\neg Q).$$

Show further that

(a) $(\neg P) \equiv P|P$

(b) $(P \,\&\, Q) \equiv (P|Q)|(Q|P)$

(c) $(P \text{ or } Q) \equiv (P|P)|(Q|Q)$

(d) $(P \Rightarrow Q) \equiv P|(Q|Q)$.

Hence deduce that any compound statement may be written using only the stroke connective.

(Remark: this may be economical in terms of connectives, but is

$$(((P|P)|Q)|((P|P)|Q))|(Q|Q)$$

easier to read than

$$((\neg P) \,\&\, Q) \Rightarrow Q?$$

They are equivalent . . .)

7

Mathematical proof

IN the last chapter we looked at the logical use of language in mathematics and the way in which the truth of a statement is deduced from given ones. We showed how a proof may be thought of as being built up from a sequence of logical deductions. In practice this does not provide a satisfactory way of *writing* proofs: to include every single step leads to a stereotyped format and is usually unbearably long-winded. In this chapter we look at the overall picture of a mathematical proof and see how such a proof is written by practising mathematicians. In addition to the underlying logical skeleton, the writing of mathematical proofs needs a sense of judgement as to how much detail is really appropriate to the given task: what must be included and what may safely be left out. Too little detail may omit vital portions of the argument; too much may obscure the overall shape.

We begin by taking an actual proof, written in normal mathematical style, and comparing it with the formal structure of the previous chapter.

THEOREM. If (a_n), (b_n) are sequences of real numbers such that $a_n \to a$ and $b_n \to b$ as $n \to \infty$, then $a_n + b_n \to a + b$.

PROOF. Let $\varepsilon > 0$. Since $a_n \to a$ there exists N_1 such that
$$n > N_1 \Rightarrow |a_n - a| < \tfrac{1}{2}\varepsilon.$$
Since $b_n \to b$ there exists N_2 such that
$$n > N_2 \Rightarrow |b_n - b| < \tfrac{1}{2}\varepsilon.$$
Let $N = \max(N_1, N_2)$. If $n > N$ then $|a_n - a| < \tfrac{1}{2}\varepsilon$ and $|b_n - b| < \tfrac{1}{2}\varepsilon$, so, by the triangle inequality,
$$|(a_n + b_n) - (a + b)| \leqslant |a_n - a| + |b_n - b|$$
$$\leqslant \tfrac{1}{2}\varepsilon + \tfrac{1}{2}\varepsilon$$
$$= \varepsilon.$$
Hence $a_n + b_n \to a + b$, as required. $\quad\square$

To analyse the structure of this proof, let us break it down line by line, adding a few words here and there to make the construction clearer.

HYPOTHESES

H_1. (a_n) is a sequence of real numbers and $a_n \to a$.
H_2. (b_n) is a sequence of real numbers and $b_n \to b$.

PROOF

1. Let $\varepsilon > 0$.
2. Since $a_n \to a$ there exists N_1 such that $n > N_1 \Rightarrow |a_n - a| < \frac{1}{2}\varepsilon$.
3. Since $b_n \to b$ there exists N_2 such that $n > N_2 \Rightarrow |b_n - b| < \frac{1}{2}\varepsilon$.
4. Let $N = \max(N_1, N_2)$.
5. If $n > N$ then $|a_n - a| < \frac{1}{2}\varepsilon$ and $|b_n - b| < \frac{1}{2}\varepsilon$.
6. So $|(a_n + b_n) - (a + b)| \leqslant |a_n - a| + |b_n - b|$ by   the   triangle inequality.
7. $|a_n - a| + |b_n - b| < \frac{1}{2}\varepsilon + \frac{1}{2}\varepsilon$.
8. $\frac{1}{2}\varepsilon + \frac{1}{2}\varepsilon = \varepsilon$.
9. (There   exists $N = \max(N_1, N_2)$ such   that) $n > N \Rightarrow$ $|(a_n + b_n) - (a + b)| < \varepsilon$.
10. $a_n + b_n \to a + b$, as required.

Line 1 is a predicate '$\varepsilon > 0$', and if we look farther down we see that eventually, in line 9, we establish the existence of N such that $n > N \Rightarrow |(a_n + b_n) - (a + b)| < \varepsilon$. Let

$$P(\varepsilon) \text{ be the predicate } \varepsilon > 0,$$

and let

$$Q(\varepsilon) \text{ be the predicate } \exists N \in \mathcal{N} : n > N \Rightarrow |(a_n + b_n) - (a + b)| < \varepsilon.$$

Then lines 1–9 are a proof of

$$P(\varepsilon) \Rightarrow Q(\varepsilon).$$

In line 10 this is translated into the equivalent statement $a_n + b_n \to a + b$.

Now let's concentrate on lines 1–9 which contain the meat of the proof. Lines 2 and 3 are translations of hypotheses H_1 and H_2, involving the implicit step that if $\varepsilon > 0$ then $\frac{1}{2}\varepsilon > 0$ so $\frac{1}{2}\varepsilon$ can be used in the definition of limit. In principle we ought to write out these short deductions explicitly, but in practice the steps are omitted when they are known parts of our technique.

Line 4 is something new: a definition of the symbol N in terms of N_1 and N_2. This definition could be omitted if we desired, and each occurrence of N in the proof replaced by $\max(N_1, N_2)$ without any real change in the proof. In practice, however, it is common to use new symbols to stand for complex concepts built up from known ones, in order to keep the notation looking simple.

Line 5 follows from lines 2, 3, 4 (although the statement that $n > N$ implies $n > N_1$ and $n > N_2$ is taken for granted).

Line 6 is arithmetic. It subsumes some relatively simple arithmetic manipulations involving rearranging $|(a_n + b_n) - (a + b)|$ to give $|(a_n - a) + (b_n - b)|$, before using the triangle inequality to give the final result. Note that this line looks like a predicate in n, but tacitly we regard it as coming under the implied quantifier $\forall n > N$ in line 5.

Line 7 follows from lines 5 and 6, again using an implicitly understood arithmetic result, this time addition of inequalities.

Line 8 is arithmetic.

Line 9 follows from lines 2–8.

This analysis of a relatively simple proof shows that mathematicians don't write out proofs precisely in the manner described in the previous chapter. Steps are omitted, both when hypotheses are introduced, and when deductions are made; new definitions are brought in; and the whole package is wrapped up in a flowing prose style in total contrast to a formal sequence of statements.

Why is this? In the first place, mathematicians were writing proofs long before they were logically analysed, so the prose style came first and continues to be used. The main reason is that the omission of trivial detail and the use of new symbols for complicated constructions are part of the process of attempting to make the deductions more comprehensible. The human mind builds up theories by recognizing familiar patterns and glossing over details which are well understood, so that it can concentrate on the new material. In fact it is limited by the amount of new information it can hold at any one time, and the suppression of familiar detail is often essential for a grasp of the total picture. In a written proof, the step by step logical deduction is therefore foreshortened where it is already a part of the reader's basic technique, so that he can comprehend the overall structure.

When working out a new theory the practising mathematician tends to distinguish between well established facts which are part of his technique, and those which are in the new material he is developing.

He then takes the established ideas very much for granted, telescoping several steps into a single line where his technique is fluent, often without giving explicit references to where a proof of these results can be found. He does this in the confidence that, if he were challenged, he would be able to fill in the details (though it might tax his memory to recall them straight away!).

Newly established results constitute the heart of the theory that is being developed, and are therefore treated with great care. They will be stated clearly as hypotheses when they are needed, and references to their proofs will be given.

When to omit logical steps or references in a proof, and when to give them in full, is part of the elusive quality: mathematical style. Different mathematicians will differ in their opinion. The clue is to look at the context of a proof and see for whom it is intended. Thus the present reader is most probably a student whose experience comes mainly from explanatory textbooks and lectures. Here the balance is likely to lean towards fuller exposition. On the other hand communication between two experts might comprise very sketchy outlines concentrating on the important new details. Nevertheless both extremes have in common the feature of omitting detail when it may be considered as part of the basic technique in the given context.

As a specific example, in studying analysis the rules of arithmetic might be subsumed as being part of the basic toolkit; but new ideas such as limits, continuity, and so on are treated with greater respect. Theorems about these new ideas are proved carefully, and later theorems refer back explicitly to results established earlier on. In the proof analysed above, arithmetic results were used without comment, though the triangle inequality was mentioned because it was felt to be sufficiently new to be worth reminding the reader about. In more advanced work it would become part of the underlying technique and be used without special reference.

In principle, a proof as used by a practising mathematician has the structure described in the previous chapter. But it occurs in a context where certain results have become a standard part of the contextual technique. A proof of a statement D from explicit hypotheses $H_1, \ldots, H_n$ will now consist of a number of statements $L_1, \ldots, L_n$, where L_n is D, and each L_m is either:

(i) A known truth, which is either a simple deduction from the hypotheses or from the contextual technique;

(ii) A deduction from the previous lines $L_1, \ldots, L_{m-1}$ using formal logic and known truths from the contextual technique.

The proof is written in a mixture of prose and mathematical symbolism which makes clear the logical structure. Steps may be omitted if the deductions are clear from the context, and new symbols may be introduced to simplify the notation.

Similarly an actual proof of a statement $P \Rightarrow Q$ will have the same underlying structure as the formal, logical one, but making tacit use of the contextual technique.

Sometimes the context may be so clear that no explicit hypotheses are mentioned. For example:

THEOREM (Euclid).   There exist infinitely many primes.

PROOF.   Suppose that there exist only a finite number of primes, say $p_1, \ldots, p_n$. The number $N = 1 + p_1 \ldots p_n$ is divisible by some prime p. But p cannot be any of $p_1, \ldots, p_n$ since the latter all leave remainder 1 on dividing N. This contradicts our assumption that $p_1, \ldots, p_n$ is the complete list of primes. $\square$

This proof depends on the context of arithmetic of whole numbers, including factorization of numbers into primes. It is given in the form of a proof by contradiction. Let P be the statement 'there exist infinitely many primes'. The first line is 'suppose $\neg P$', and the proof thereafter follows the usual line of argument in search of a contradiction. Then $\neg P$ must be false, so P must be true.

In a proof like this we must keep a careful eye on our contextual material for the presence of logical flaws in the parts of the proof which have been omitted, as well as on the actual symbols on the paper. For instance, what is wrong with the following 'proof'?

THEOREM(?).   1 is the largest integer.

PROOF(?).   Suppose not. Let n be the largest integer. Then $n > 1$. Now n^2 is also an integer, and $n^2 > n.1 = n$. So $n^2 > n$, which contradicts n being the largest integer.

Therefore our initial assumption is false, and 1 is the largest integer, as claimed. $\square$

Where is the flaw? Think about it before reading on.

The flaw is in the statement 'Let n be the largest integer'. This is not the correct negation of '1 is the largest integer'. It should be '$n > 1$ is the largest integer *or there is no largest integer*', and with this statement substituted the contradiction fails to materialize, since $n^2 > n$ does not contradict the phrase italicized above.

This logical flaw is disguised by the informality of the proof. It needs a lot of experience to avoid traps like these.

Axiomatic systems

To make sure that the contextual material used has a firm basis, we must start somewhere. We do this by taking certain explicit statements as basic *axioms*, which are assumed to be true, and from these all other results in the theory are deduced. According to taste, these deductions are called theorems, propositions, lemmas, corollaries, and so on. The words 'theorem' and 'proposition' are often regarded as being interchangeable, some authors sticking exclusively to one or the other of these. We prefer to use the word 'proposition' to describe ordinary run-of-the-mill results, reserving the word 'theorem' for something more important. In this way the structure of the theory can be seen more clearly, with important theorems standing out in relief from the background of propositions.

To give even more shape to the contours of the theory and to reduce the strain on particularly long proofs, constituent parts of a proof may be separated out and proved before a given theorem or proposition. Such a preliminary result is called a *lemma*. There may be several lemmas preceding a major theorem, so that when the proof of the main result is reached, all the spadework has been done, and all that is left is to put the pieces together. In this way it is possible to make the proof of the theorem itself a much more streamlined affair, with its salient features clearly delineated and not concealed by the details which have been subordinated in the lemmas.

The complement of a lemma (which precedes a theorem) is a *corollary*. This is a result which can be deduced very simply from a theorem (or proposition) and immediately follows it. Sometimes the proof is so obvious, because of the context, that the proof of a corollary is omitted, or 'left to the reader'.

By grading the titles of results proved as theorems, propositions, lemmas, and corollaries it is possible to give a more coherent shape to the theory as a whole and make it more accessible. Meanwhile, as new levels of sophistication are reached, it is possible to assume more results as basic contextual material in the manner described earlier.

In chapter 2 we looked at intuitive ideas of the real numbers and proved results in that context. To treat the subject formally, we will have to select certain properties of arithmetic as basic axioms and then build logically on these. (If we are sensible, we will use all the guile we have developed in intuitive arithmetic to suggest to us which way we should go in our formal development.) In chapter 8 we will look at suitable axioms for the natural numbers before moving on to other number systems. Once we have a firm foundation for arithmetic, we can use it as contextual material to go on to more advanced theories. In handling vector spaces, or analysis, or geometry, the arithmetic results can be subsumed and we can concentrate on the next level of deduction. At each stage it should be made clear what type of result may be used without comment, and which must be documented carefully. Sometimes the author of a textbook, or a lecturer, may assume that the context makes this clear, and fail to mention it explicitly.

Examination questions

There is one situation which is of great concern to students, and that is what constitutes a proof acceptable in an examination. To a certain extent this depends on the examiner, but part of the anxiety is due to ignorance of the context. In a book, the context is clear from the position. A proof in chapter 7 is obviously allowed to assume results from chapters 1–6. But in an examination it may not be clear at which level a proof is required. Do all the steps have to be included? What can safely be missed out? If the question is well posed, it will make the context clear. The phrase 'show from first principles . . .' obviously asks for a careful proof from the basic definitions and axioms. A question on the more advanced parts of a subject will not expect this kind of answer, and it is safe to assume any preceding material which is well established as contextual material for that level, never going into greater detail than that which is appropriate to the concepts used in the question. In particular, if a question is asked in a manner which makes familiar use of certain ideas, then they can be used at the same level of familiarity in the solution. This prevents long-winded answers including proofs of elementary material which ought to be subsumed in the context.

Levels may vary within a single question, with the first part being elementary and later parts more advanced. The wise student will sensibly increase the power of his reasoning to the appropriate context, freely using ideas commensurate with the new situation.

Exercises

1. Is the following a proof? If not, why not?
Theorem: For all real numbers x, y,
$$\tfrac{1}{2}(x+y) \geqslant \sqrt{xy}.$$

Proof: Squaring and multiplying through by 4,

$$x^2 + 2xy + y^2 \geqslant 4xy$$

so subtracting $4xy$ from each side,

$$x^2 - 2xy + y^2 \geqslant 0.$$

But $x^2 - 2xy + y^2 = (x-y)^2$ which is always $\geqslant 0$, so the theorem is proved.

2. Is the following a proof? If not, why not?

Theorem: The base angles of an isosceles triangle are equal.

Proof: Let $\triangle ABC$ be an isosceles triangle with sides $AB = AC$. Then $\triangle ABC$ is congruent to $\triangle ACB$, because the corresponding sides are equal: $AB = AC$, $BC = CB$, $AC = AB$. Here corresponding angles are equal: in particular $\angle ABC = \angle ACB$.

(You may assume the usual geometrical properties of congruent triangles.)

3. Are the 'proofs' given in chapter 2 of this book genuine proofs within a suitable context? If so, what is the context? If not, what are the 'proofs'?

4. Analyse the proof of proposition 5, chapter 3, showing how each statement follows from previous ones. What must be added to the proof as written to make it fit the logical definition of a 'proof'?

Repeat the exercise for other proofs from chapter 3.

5. Find a mathematics textbook, select a theorem (whose proof is neither too long nor too short) and analyse its structure. What results are assumed as contextual background?

Repeat the exercise for several other theorems, preferably from different texts and in different branches of mathematics.

6. The following are axioms for a (hitherto undefined) mathematical structure known as a *bureaucracy*. This consists of

a set B of *bureaucrats*,

a set C of *committees*,

a relation S between B and C (read *serves on*),

satisfying the following axioms:

(B1) Every bureaucrat serves on at least three different committees.

(B2) Every committee is served on by at least three different bureaucrats.

(B3) Given two distinct committees, exactly one bureaucrat serves on both.

(B4) Given two distinct bureaucrats, there is exactly one committee on which they both serve.

Prove from these axioms that if the number of bureaucrats is finite, so is the number of committees. Prove that there are always at least seven bureaucrats in a bureaucracy, and find a bureaucracy with exactly seven bureaucrats.

7. The following proof fits the logical definition. Analyse it to find out what is really going on.

Theorem: If A, B, C are sets then $(A \cap B) \cap C = A \cap (B \cap C)$.

Proof:

L_1: Let $a \in (A \cap B) \cap C$.

L_2: $a \in A \cap B$.

L_3: Let $b \in A \cap (B \cap C)$.

L_4: $a \in C$.

L_5: $b \in B \cap C$.

L_6: $b \in B$.

L_7: $a \in B$.

L_8: $b \in C$.

L_9: $\{a, b\} \subseteq B$.

L_{10}: $b \in A$.

L_{11}: $a \in A$.

L_{12}: $b \in A \cap B$.

L_{13}: $a \in A \cap B$.

L_{14}: $\{a, b\} \subseteq A \cap B$.

L_{15}: $a \in B \cap C$.

L_{16}: $a \in A \cap (B \cap C)$.

L_{17}: $(A \cap B) \cap C \subseteq A \cap (B \cap C)$.

L_{18}: $b \in (A \cap B) \cap C$.

L_{19}: $(A \cap B) \cap C \supseteq A \cap (B \cap C)$.

L_{20}: $(A \cap B) \cap C = A \cap (B \cap C)$.

Rewrite it in a sensible style so that the structure of the argument becomes apparent.

PART III

The development of axiomatic systems

Now we turn to the number systems themselves, analysing their structure, and aiming to find a formal list of axioms that will describe them precisely. We also show how to construct systems which satisfy these axioms using the raw materials of set theory. This places our intuitive ideas on a firm basis and allows us to make use of them without any logical qualms.

Metaphorically, we are now constructing our building, or growing our plant: the important thing is to take as much care as is required to make sure that nothing goes wrong, and this means a certain amount of attention to detail.

The attitude of mind demanded of the reader is now a little different. Although intuitive ideas may be used as a source of inspiration, nothing may be used as part of a proof until it has been given a rigorous logical demonstration. It therefore becomes necessary to give rigorous proofs from the axioms of properties which, on an intuitive level, we already accept. We do this in order to be sure that, in this axiomatic sense, they really are true, for our previous acceptance of them is based more on habit than on logic. We thus put our ideas on a sounder basis.

In chapter 8 we give highly detailed proofs, even of statements that may seem obvious. However from chapter 9 onwards we gradually relax the amount of detail, leaving parts of the proof which by now have become routine for the reader to check for himself. This is essential to avoid losing track of the main outline beneath an accumulation of ever more elaborate detail. The step-by-step method, if carried too far, obscures the overall picture.

8

Natural numbers and proof by induction

AT first sight a proof by induction doesn't seem to fit the pattern of proof described in the last chapter. Let us look at a typical instance:

PROPOSITION. The sum of the first n natural numbers is $\frac{1}{2}n(n+1)$.

PROOF. This is trivially true for $n = 1$. If it is true for $n = k$,

$$1+2+ \ldots +k = \tfrac{1}{2}k(k+1),$$

then adding $k+1$ to each side we obtain

$$1+2+ \ldots +(k+1) = \tfrac{1}{2}k(k+1)+(k+1) = \tfrac{1}{2}(k+1)(k+2).$$

This is the sum of the first $k+1$ natural numbers and the formula is true for $n = k+1$. By induction, the formula is true for all natural numbers. □

Many people regard this as an 'and so on . . .' sort of argument in which the truth of the statement is established for $n = 1$; then, having established the general step from $n = k$ to $n = k+1$, this is applied for $k = 1$ to get us from $n = 1$ to $n = 2$, then used again to go from $n = 2$ to $n = 3$, and so on, as far as we wish to go. For instance we could reach $n = 593$ after 592 applications of the general step. The only trouble is that to reach large values of n requires a large number of applications of the general step, and we can never actually cover *all* natural numbers in a single proof of finite length.

The way out of this dilemma is to remove the 'and so on . . .' part from the proof and place it squarely in the actual definition of the natural numbers. Proof by induction will then be seen to fit into the type of mathematical proof described in the last chapter.

The natural numbers

The natural numbers form a highly non-trivial set, because we cannot write down a complete list of elements: they just go on for ever. To describe them satisfactorily needs a different approach. Fortunately the intuitive idea of counting can easily be modelled in a set-theoretic way. We begin with 1, then comes 2, then 3, and we carry on in this way naming each successive number as far as we wish. To grasp the concept of the set of natural numbers 'all in one go' we regard this succession as a *function* on the set $\mathcal{N}$ of natural numbers. That is, we seek a function $s: \mathcal{N} \to \mathcal{N}$ with suitable properties. Here s stands for 'successor' and $s(1) = 2$, $s(2) = 3$, etc. Two obvious properties that we need are:

(i) s is not surjective (because $s(n) \neq 1$ for any $n \in \mathcal{N}$),

(ii) s is injective ($s(m) = s(n)$ implies $m = n$).

There is a third vital property, giving rise to induction proofs, as follows:

(iii) Suppose that $S \subseteq \mathcal{N}$ is such that $1 \in S$; and for all $n \in \mathcal{N}$ if $n \in S$ then $s(n) \in S$. Then $S = \mathcal{N}$.

In words, (iii) says that a subset containing 1 which includes $s(n)$ whenever it contains n, exhausts the whole set of natural numbers.

It is a surprising fact that these three properties are all that is required to describe the natural numbers. An axiomatic basis for arithmetic requires only that we postulate the existence of a set with these three properties. For technical reasons it is more profitable to start with 0 rather than 1. The reasons are not hard to find. Although in counting we usually start with 1, the empty set has 0 elements and it is useful to be able to say so. Again, in arithmetic it is convenient to have the zero element. For these and other reasons we shall start with 0 in our axiomatic system, and to avoid confusion with our intuitive concept of the natural numbers we shall use $\mathbb{N}_0$ to denote the formal system, the suffix serving as a reminder that 0 is included. We then obtain the *Peano axioms* for the natural numbers, named after the Italian mathematician responsible for this approach at the end of the nineteenth century:

We suppose that there exists a set $\mathbb{N}_0$ and a function $s: \mathbb{N}_0 \to \mathbb{N}_0$ such that

(N1) s is not surjective: there exists $0 \in \mathbb{N}_0$ such that $s(n) \neq 0$ for any $n \in \mathbb{N}_0$.

(N2) s is injective: if $s(m) = s(n)$ then $m = n$.

(N3) If $S \subseteq \mathbb{N}_0$ is such that $0 \in S$; and $n \in S \Rightarrow s(n) \in S$ for all $n \in \mathbb{N}_0$, then $S = \mathbb{N}_0$.

There is no guarantee that any such set $\mathbb{N}_0$ exists, so we must take its existence as a basic axiom for our mathematics:

Existence axiom for the natural numbers.   There exists a set $\mathbb{N}_0$ and a function $s: \mathbb{N}_0 \to \mathbb{N}_0$ satisfying (N1)–(N3).

From these slender beginnings we can develop all the usual properties of arithmetic, then later build up the other number systems, including the real and complex numbers. We will also see how axiom (N3) enshrines the idea of proof by induction, as in the following simple case:

PROPOSITION 1.   If $n \in \mathbb{N}_0$, $n \neq 0$, then there exists a unique $m \in \mathbb{N}_0$ such that $n = s(m)$.

PROOF.   Let $S = \{n \in \mathbb{N}_0 | n = 0 \text{ or } n = s(m) \text{ for some } m \in \mathbb{N}_0\}$. Certainly $0 \in S$. If $n \in S$ then either $n = 0$, in which case $s(n) = s(0)$ so $s(n) \in S$; or $n = s(m)$ and $s(n) = s(s(m))$ where $s(m) \in \mathbb{N}_0$, so $s(n) \in S$. Hence by axiom (N3), $S = \mathbb{N}_0$. This shows that the required m exists. Uniqueness follows from (N2).   □

Proposition 1 tells us that 0 is the only element which is not a successor, a property which distinguishes it from all other elements. The set $\mathbb{N} = \mathbb{N}_0 \backslash \{0\}$ will be called the *natural numbers*. We shall denote $s(0)$ by 1. This element lies in $\mathbb{N}$ and will prove of paramount importance.

Look at the proof of proposition 1 once more. Its essential structure consists of defining a set S, then

   (i) showing $0 \in S$,

  (ii) showing that $n \in S \Rightarrow s(n) \in S$,

 (iii) invoking axiom (N3) to deduce that $S = \mathbb{N}_0$.

A proof by induction always has this format.

In practice S is of the form

$$S = \{n \in \mathbb{N}_0 | P(n)\}$$

where $P(n)$ is a predicate known to be true or false for each $n \in \mathbb{N}_0$. The statements (i), (ii), (iii) translate into

   (i)' showing $P(0)$ is true,

  (ii)' showing that if $P(n)$ is true then $P(s(n))$ is true,

 (iii)' invoking (N3) to deduce that $P(n)$ is true for all $n \in \mathbb{N}_0$.

Using axiom (N3) finishes the proof without a breath of an 'and so on . . .' type of argument. The reader will recognize the basic skeleton of this method in the proposition which started the chapter, except that we started at 1 instead of 0 and wrote $n + 1$ instead of $s(n)$. Later on we will show that the same method applies starting at any $k \in \mathbb{N}_0$, in particular at $k = 1$, so the proposition at the beginning of the chapter is just a simple example of an induction proof depending on the use of axiom (N3).

In practice axiom (N3) may not be mentioned explicitly. The proof may be phrased entirely in terms of a predicate $P(n)$, and when steps (i)' and (ii)' are established, then the conclusion '$P(n)$ is true for all n' is said to be established 'by induction'. The reader should always interpret this as an implicit use of axiom (N3), which is referred to as the *induction axiom* for this very reason. During the course of such a proof, the assumption that $P(n)$ is true is called the *induction hypothesis* and the proof that $P(n) \Rightarrow P(s(n))$ is called the *induction step*. For the moment, as we begin to grapple with $\mathbb{N}_0$, we shall always make the set S explicit.

Definition by induction

The most important matter to get to grips with is the arithmetic of $\mathbb{N}_0$. To start we must define the basic operations of addition and multiplication. We can define addition by setting

$$m + 0 = m \tag{1}$$

for all $m \in \mathbb{N}_0$, and then defining $m + s(n)$ in terms of $m + n$ by

$$m + s(n) = s(m + n). \tag{2}$$

The induction axiom seems tailor-made for definitions, as well as proofs. But there is one flaw that we must eliminate, arising from a fundamental difference between proof and definition. In an induction proof, the induction step $n \in S \Rightarrow s(n) \in S$ only involves showing that *if* $n \in S$ is true, then so is $s(n) \in S$; we don't actually have to know whether $n \in S$ is true or not. But in making an inductive definition of addition, to be able to define $m + s(n)$ by (2) it is absolutely vital to know the value of $m + n$ first. Of course, our intuitive model $\mathcal{N}$ tells us that starting from (1) we can use step (2) n times to find $m + n$. Unfortunately in $\mathbb{N}$ we have not established any such principle; indeed given $m \in \mathbb{N}$ we don't yet know that starting with 0 and forming successors $1 = s(0)$, $2 = s(1)$, and so on, we eventually reach m. Indeed our intention is to eliminate 'and so on . . .' arguments. The way to remove this flaw is to prove a general principle on the validity of such definitions. The one we shall prove covers all the cases arising in this chapter, although more general ones are possible.

THEOREM 2 (*Recursion Theorem*).   If X is a set, $f: X \to X$ a function, and $c \in X$, then there exists a unique function $\phi: \mathbb{N}_0 \to X$ such that

(i) $\phi(0) = c$,

(ii) $\phi(s(n)) = f(\phi(n))$ for all $n \in \mathbb{N}_0$.

PRE-PROOF DISCUSSION.   In terms of the set-theoretic notation in chapter 5 a function $\phi: \mathbb{N}_0 \to X$ is a subset of $\mathbb{N}_0 \times X$ with the properties

(I) Given $n \in \mathbb{N}_0$ there exists $x \in X$ such that $(n, x) \in \phi$,

(II) If (n, x) and $(n, y) \in \phi$ then $x = y$.

So to define ϕ it suffices to specify the appropriate subset. Now, translating the statement of the theorem, this subset must satisfy

(a) $(0, c) \in \phi$,

(b) $(n, x) \in \phi \Rightarrow (s(n), f(x)) \in \phi$.

Now, there exist many subsets with these properties, such as $\mathbb{N}_0 \times X$ itself. The one we want is the intersection of all such subsets, since only this satisfies (I) and (II).

PROOF.   Let ϕ be the intersection of all subsets U of $\mathbb{N}_0 \times X$ satisfying

$$(0, c) \in U, \tag{3}$$

$$(n, x) \in U \Rightarrow (s(n), f(x)) \in U. \tag{4}$$

Then ϕ will satisfy both of these conditions; furthermore, it will be the *smallest* set which does so. It remains to show that ϕ really is a function, which means verifying (I) and (II).

Now (I) is trivially established by letting

$$S = \{n \in \mathbb{N}_0 | (n, x) \in \phi \text{ for some } x \in X\},$$

for (a) says that $0 \in S$ and (b) that $n \in S \Rightarrow s(n) \in S$. By induction $S = \mathbb{N}_0$.

(II) is a little trickier (but not much). Let

$$T = \{n \in \mathbb{N}_0 | (n, x) \in \phi \text{ for a unique } x \in X\}.$$

We know that $(0, c) \in \phi$. If also $(0, d) \in \phi$ with $c \neq d$, let

$$\phi^- = \phi \setminus \{(0, d)\}.$$

Then ϕ^- satisfies (3); and if $(n, x) \in \phi^-$ then $(s(n), f(x)) \in \phi$ and is not $(0, d)$ because $s(n) \neq 0$ by axiom (N1). So $(s(n), f(x)) \in \phi^-$ and ϕ^- satisfies (4). Since ϕ is the smallest set satisfying (3) and (4) this is a contradiction, hence no such d exists, so $0 \in T$.

The induction step follows the same pattern. If $n \in T$ then $(n, x) \in \phi$ for precisely one $x \in X$. From (b) we have $(s(n), f(x)) \in \phi$, so to establish that $s(n) \in T$ we must show that no other ordered pair $(s(n), y) \in \phi$ with $y \neq f(x)$. If there were such a pair, consider $\phi^* = \phi \setminus \{(s(n), y)\}$. Again, since $0 \neq s(n)$, we know that ϕ^* satisfies (3). To check (4) we need to prove that

$$(m, z) \in \phi^* \Rightarrow (s(m), f(z)) \in \phi^* \quad \text{for all } m \in \mathbb{N}_0.$$

Now for $m = n$ this is true, since we have only one $x \in X$ such that $(n, x) \in \phi$, and for this x $(s(n), f(x)) \in \phi$ by (b) and is not $(s(n), y)$ since $y \neq f(x)$. For $m \neq n$ we have $(s(m), f(z)) \in \phi$ by (b), and $s(m) \neq s(n)$ by (N2). Hence $(s(m), f(x)) \neq (s(n), y)$, and it follows that $(s(m), f(z)) \in \phi^*$. Either way, ϕ^* satisfies (4) and again we have a contradiction. By induction $T = \mathbb{N}_0$, which completes the proof. $\square$

As examples of the recursion theorem in action we have:
(1) *Addition.* $\phi_m : \mathbb{N}_0 \to \mathbb{N}_0$, $\phi_m(n) = m + n$, defined by

$$\phi_m(0) = m$$

$$\phi_m(s(n)) = s(\phi_m(n)).$$

Here $c = m, f = s$.
(2) *Multiplication.* $\mu_m : \mathbb{N}_0 \to \mathbb{N}_0$, $\mu_m(n) = mn$, defined by

$$\mu_m(0) = 0$$

$$\mu_m(s(n)) = \mu_m(n) + m.$$

Here $c = 0, f(r) = r + m$.
(3) *Powers.* $\pi_m : \mathbb{N}_0 \to \mathbb{N}_0$, $\pi_m(n) = m^n$, defined by
$$\pi_m(0) = 1$$

$$\pi_m(s(n)) = m\pi_m(n).$$

Here $c = 1, f(r) = rm$.
(4) *Repeated composition* of a map $f : X \to X$, defined by

$$f^0(x) = x$$

$$f^{s(n)}(x) = f(f^n(x)) \quad \text{for all } x \in X.$$

The rules of arithmetic

With addition and multiplication properly defined, it is now a relatively easy matter to prove the usual rules of arithmetic. The proofs are not absolutely trivial, as the reader may find if he tries to do the job himself without any guidance. The main tool is induction, the main problem is to find the best order in which to prove things. Conceivably a talented reader might improve on the presentation given below.

For reference we recall the definitions:

$$(\alpha \mathrm{i})\ m+0=m, \qquad (\alpha \mathrm{ii})\ m+s(n)=s(m+n),$$

$$(\mu \mathrm{i})\ m0=0, \qquad (\mu \mathrm{ii})\ ms(n)=mn+m.$$

Now from $(\alpha \mathrm{ii})$ and $(\alpha \mathrm{i})$ we see that $m+s(0)=s(m+0)=s(m)$. We have already denoted $s(0)$ by 1, so $s(m)=m+1$.

LEMMA 3.   For all $m \in \mathbb{N}_0$,
(a) $0+m = m$,
(b) $1+m = s(m)$,
(c) $0m = 0$,
(d) $1m = m$.

PROOF.   In each case we use induction on m. We shall verify (a) and leave the rest as an exercise. Let

$$S = \{m \in \mathbb{N}_0 | 0+m = m\}.$$

Trivially $0 \in S$ by $(\alpha \mathrm{i})$. Now if $m \in S$ then $0+m = m$, so that by $(\alpha \mathrm{ii})$ $0+s(m)=s(0+m)=s(m)$. Therefore $s(m) \in S$. By (N3), $S = \mathbb{N}_0$.

THEOREM 4.   For all $m, n, p \in \mathbb{N}_0$,
(a) $(m+n)+p = m+(n+p)$
(b) $m+n = n+m$
(c) $(mn)p = m(np)$
(d) $mn = nm$
(e) $m(n+p) = mn + mp$.

PROOF.   (a) is by induction on p, using

$$S = \{p \in \mathbb{N}_0 | (m+n)+p = m+(n+p)\}.$$

First

$$(m+n)+0 = m+n \qquad \text{by } (\alpha \mathrm{i})$$

$$= m+(n+0) \quad \text{by } (\alpha \mathrm{i}),$$

so $0 \in S$.

Second, if $p \in S$ then

$$(m+n)+p = m+(n+p), \tag{5}$$

so

$$(m+n)+s(p) = s((m+n)+p) \quad \text{by } (\alpha\text{ii})$$
$$= s(m+(n+p)) \quad \text{by } (5)$$
$$= m+s(n+p) \quad \text{by } (\alpha\text{ii})$$
$$= m+(n+s(p)) \quad \text{by } (\alpha\text{ii})$$

implying $s(p) \in S$. By induction $S = \mathbb{N}_0$.

(b) is by induction on n using

$$S = \{n \in \mathbb{N}_0 | m+n = n+m\}.$$

Lemma 3(a) shows $0 \in S$. If $n \in S$, then

$$m+n = n+m \tag{6}$$

and then

$$m+s(n) = s(m+n) \quad \text{by } (\alpha\text{ii})$$
$$= s(n+m) \quad \text{by } (6)$$
$$= n+s(m) \quad \text{by } (\alpha\text{ii})$$
$$= n+(1+m) \quad \text{by lemma } 3(b)$$
$$= (n+1)+m \quad \text{by theorem } 4(a)$$
$$= s(n)+m,$$

hence $s(n) \in S$. By induction $S = \mathbb{N}_0$, establishing (b).

It is convenient to deal with (e) next, using induction on p. Let

$$S = \{p \in \mathbb{N}_0 | m(n+p) = mn + mp\}.$$

Then

$$m(n+0) = mn \quad \text{by } (\alpha\text{i})$$
$$= mn + 0 \quad \text{by } (\alpha\text{i})$$
$$= mn + m0 \quad \text{by } (\mu\text{i}),$$

implying $0 \in S$.

If $p \in S$, then

$$m(n+p) = mn + mp \tag{7}$$

and hence

$$m(n+s(p)) = ms(n+p) \qquad \text{by } (\alpha\text{ii})$$
$$= m(n+p)+m \qquad \text{by } (\mu\text{ii})$$
$$= (mn+mp)+m \quad \text{by } (7)$$
$$= mn+(mp+m) \quad \text{by } (a)$$
$$= mn+ms(p) \qquad \text{by } (\mu\text{ii}),$$

so $s(p) \in S$ and induction gives $S = \mathbb{N}_0$.

The proof of (c) is now relatively straightforward and of the same nature as previous proofs. This leaves (d), which turns out to be a little more tricky. Let

$$S = \{n \in \mathbb{N}_0 | mn = nm\}.$$

Now $0 \in S$ by lemma 3(c). If $n \in S$ then

$$mn = nm \qquad (8)$$

and we have

$$ms(n) = mn + m \quad \text{by } (\mu\text{ii})$$
$$= nm + m \quad \text{by } (8).$$

If we could show that this equalled $s(n)m$ we would have finished, but unfortunately we don't know this yet. However, we can prove it by a second induction on m. Let

$$T = \{m \in \mathbb{N}_0 | nm + m = s(n)m\}.$$

Then $0 \in T$, and if $m \in T$ then

$$nm + m = s(n)m \qquad (9)$$

and hence

$$ns(m) + s(m) = n(m+1) + (m+1)$$
$$= (nm+n) + (m+1) \quad \text{by } (e)$$
$$= nm + (n + (m+1)) \quad \text{by } (a)$$
$$= nm + ((n+m)+1) \quad \text{by } (a)$$
$$= nm + ((m+n)+1) \quad \text{by } (b)$$
$$= nm + (m + (n+1)) \quad \text{by } (a)$$

$$= (nm + m) + (n + 1) \quad \text{by (a)}$$

$$= s(n)m + s(n) \qquad \text{by (9)}$$

$$= s(n)s(m) \qquad \text{by } (\mu\text{ii}),$$

hence $s(m) \in T$ and $T = \mathbb{N}_0$. Hence, returning to where we left off, $s(n) \in S$ and $S = \mathbb{N}_0$. This proves (d). $\square$

Having performed this massive induction exercise we can now use these arithmetic results freely. In the first place, we replace $s(n)$ by $n + 1$. The induction axiom takes on the more familiar form:

If $S \subseteq \mathbb{N}_0$, $0 \in S$, and $n \in S \Rightarrow n + 1 \in S$, then $S = \mathbb{N}_0$.

Axiom (N2) translates into

$$m + 1 = n + 1 \Rightarrow m = n,$$

and this can be extended by induction to give

PROPOSITION 5.   For all $m, n, q \in \mathbb{N}_0$,
(a) $m + q = n + q \Rightarrow m = n$
(b) $q \neq 0, mq = nq \Rightarrow m = n$.

PROOF.   (a) We use induction on q. Let

$$S = \{q \in \mathbb{N}_0 \mid m + q = n + q \Rightarrow m = n\}.$$

Trivially $0 \in S$. If $q \in S$, suppose that

$$m + (q + 1) = n + (q + 1).$$

Then by theorem 4,

$$(m + q) + 1 = (n + q) + 1,$$

hence by (N2)

$$m + q = n + q$$

and since $q \in S$,

$$m = n.$$

Hence $q + 1 \in S$ and by induction $S = \mathbb{N}_0$.
(b) Let

$$S = \{m \in \mathbb{N}_0 \mid q \neq 0, mq = nq \Rightarrow m = n\}.$$

To show $0 \in S$, suppose that $q \neq 0$, and

$$nq = 0q = 0.$$

Then $q = p + 1$ for some p. If $n \neq 0$ then $n = r + 1$. Then $nq = (pr + p + r) + 1$ so cannot be 0. Hence $n = 0$, so $0 \in S$.

Now suppose $m \in S$, and $q \neq 0$, with

$$(m + 1)q = nq.$$

As before, $n \neq 0$, so $n = r + 1$ for some $r \in \mathbb{N}_0$. Then $mq + q = rq + q$. By part (a), $mq = rq$; by hypothesis $m = r$. Therefore $m + 1 = n$. This completes the proof. $\square$

We may now discuss subtraction. Suppose that $p = r + q$. By proposition 5, r is given uniquely by p and q. We may therefore denote r by $p - q$. For $m, n \in \mathbb{N}_0$ we define the relation $\geqslant$ by

$$m \geqslant n \Leftrightarrow \exists r \in \mathbb{N}_0, \; m = r + n.$$

Given $m, n \in \mathbb{N}_0$, the difference $m - n$ is defined only when $m \geqslant n$. This being so, we can verify various rules of subtraction as required, such as

$$m - (n - r) = (m - n) + r \quad \text{for } m \geqslant n \geqslant r,$$
$$m + (n - r) = (m + n) - r \quad \text{for } n \geqslant r,$$
$$m(n - r) = mn - mr \quad \text{for } n \geqslant r.$$

All are routine; for instance the last follows by considering

$$n = s + r \quad (\text{since } n \geqslant r),$$

whence

$$mn = m(s + r) = ms + mr.$$

Thus by definition,

$$mn - mr = ms = m(n - r)$$

since $s = n - r$.

We may also consider division, and in the case $m = rn$ ($n \neq 0$) we may denote r by m/n. We shall look into the question of when division is possible in a later section.

Ordering the natural numbers

We have already defined the relation $\geq$ on $\mathbb{N}_0$. The other order relations are given by

$$m > n \Leftrightarrow m \geq n \ \& \ m \neq n,$$

$$m \leq n \Leftrightarrow n \geq m,$$

$$m < n \Leftrightarrow n > m.$$

We must prove that these are indeed order relations in the sense of chapter 4. For example,

PROPOSITION 6. $m \geq n, n \geq p \Rightarrow m \geq p$ for all $m, n, p \in \mathbb{N}_0$.

PROOF.   We have $r, s \in \mathbb{N}_0$ such that $m = r + n, n = s + p$. Hence $m = r + (s + p) = (r + s) + p$, so $m \geq p$.   □

A second property of order relations is also easy:

PROPOSITION 7.   If $m, n \in \mathbb{N}_0$ and $m \geq n, n \geq m$, then $m = n$.

PROOF.   There exist $r, t \in \mathbb{N}_0$ such that $m = r + n, n = t + m$, so $m = r + t + m$. By proposition 5(a), $r + t = 0$. We cannot have $t \neq 0$, since this would imply $t = q + 1$ for some $q \in \mathbb{N}_0$ by lemma 1, and then $0 = (r + q) + 1$, contradicting axiom (N1). Therefore $t = 0$, so $n = m$.   □

The third property of an order relation requires a more technical proof, which is postponed until proposition 12. However, it is a simple matter so see that the relations behave as expected relative to the arithmetical operations of $\mathbb{N}_0$:

PROPOSITION 8.   For all $m, n, p, q \in \mathbb{N}_0$,
(a) If $m \geq n, p \geq q$, then $m + p \geq n + q$,
(b) If $m \geq n, p \geq q$, then $mp \geq nq$.

PROOF.   (a) there exist $r, s \in \mathbb{N}_0$ such that $m = r + n, p = s + q$. Hence, after simplification, we find that $m + p = (r + s) + (n + q)$.
(b) Similarly $mp = nq + (rs + ns + rq)$.   □

The zero element 0 is the smallest element of $\mathbb{N}_0$, in the following sense:

LEMMA 9.   If $m \in \mathbb{N}_0$ then $m \geq 0$.

PROOF. $m = 0 + m$.   □

The element 1 is the next smallest:

LEMMA 10.   If $m \in \mathbb{N}_0$ and $m > 0$ then $m \geq 1$.

PROOF.   By proposition 1, if $m \neq 0$ then $m = q + 1$ for some $q \in \mathbb{N}_0$. Hence $m \geq 1$.   □

We could go on to show that $2 = 1 + 1$ is the next smallest after 1, then $3 = 2 + 1$ is the next smallest after 2, and so on. It is more efficient to prove a general proposition:

PROPOSITION 11.   If $m, n \in \mathbb{N}_0$ and $m > n$ then $m \geq n + 1$.

PROOF.   We have $m = n + r$ for some $r \in \mathbb{N}_0$, and $r \neq 0$ since $m \neq n$. By proposition 1, $r = q + 1$ for some $q \in \mathbb{N}_0$, hence $m = (n + 1) + q$, and $m \geq n + 1$.   □

Now we can complete the proof that $\geq$ is an order relation in the sense of chapter 5.

PROPOSITION 12.   The relation $\geq$ is a (weak) order relation on $\mathbb{N}_0$.

PROOF.   By definition, what we must prove is that for all $m, n, p \in \mathbb{N}_0$,

(WO1) $m \geq n$ & $n \geq p \Rightarrow m \geq p$,
(WO2) Either $m \geq n$ or $n \geq m$,
(WO3) If $m \geq n$ and $n \geq m$ then $m = n$.

We have already established (WO1) and (WO3) in propositions 6 and 7. To verify (WO2), let

$$S(m) = \{n \in \mathbb{N}_0 \mid m \geq n \text{ or } n \geq m\}.$$

We aim to prove that $S(m) = \mathbb{N}_0$ for all $m \in \mathbb{N}_0$. Now for a given m, we have $0 \in S(m)$ since $m \geq 0$. Next suppose that $n \in S(m)$. Either $m \geq n$ or $n \geq m$. If $n \geq m$ then $n + 1 \geq m$. If $n \leq m$ then either $n = m$, and $m \leq n + 1$, or $m > n$, in which case $m \geq n + 1$ by proposition 11. Thus $n + 1 \in S(m)$. By induction $S(m) = \mathbb{N}_0$, as required.   □

As remarked in chapter 4, it follows that $>$ is a strict order relation. That is, for all $m, n, p \in \mathbb{N}_0$,

$$m > n \text{ & } n > p \Rightarrow m > p,$$

Exactly one of $m > n$, $m = n$, $m < n$ is true (trichotomy law). The next result is almost a converse to proposition 5.

PROPOSITION 13.   For all $m, n, p, q \in \mathbb{N}_0$,
(a) $m + q > n + q \Rightarrow m > n$,
(b) $q \neq 0, mq > nq \Rightarrow m > n$.

PROOF.   (a) if $m \not> n$ then $m \leq n$ by trichotomy. But $m \leq n$ implies $m + q \leq n + q$ by proposition 8(a). This contradicts the hypothesis, so part (a) is proved. Part (b) follows a similar format.   □

Proposition 13 is of course valid when $>$ is replaced by $\geq$, and is then an exact converse to proposition 5.

The uniqueness of $\mathbb{N}_0$

The set $\mathbb{N}_0$, its arithmetic, and order, are essentially unique in a very precise sense. As a down-to-earth illustration, the French counting system 'un, deux, trois, . . . ', while undeniably *different* from the English 'one, two, three, . . . ', possesses the same arithmetical structure. To see this, we observe that translating from French to English by replacing 'un', by 'one', 'deux' by 'two', and so on, turns valid French arithmetic into valid English arithmetic, and conversely. It is the same with $\mathbb{N}_0$. Suppose that we can find another set $\mathbb{N}_0'$ with a function $s': \mathbb{N}_0' \to \mathbb{N}_0'$ satisfying the corresponding axioms (N'1)–(N'3). Then we define $\phi: \mathbb{N}_0 \to \mathbb{N}_0'$ by

$$\phi(0) = 0'$$

$$\phi(s(n)) = s'(\phi(n))$$

for all $n \in \mathbb{N}_0$. This function exists by the recursion theorem, as does $\varphi: \mathbb{N}_0' \to \mathbb{N}_0$ given by

$$\varphi(0') = 0$$

$$\varphi(s'(m)) = s(\varphi(m))$$

for all $m \in \mathbb{N}_0'$.

A simple induction proof shows that ϕ and φ are mutual inverses, letting $S = \{n \in \mathbb{N}_0 | \varphi\phi(n) = n\}$ to show that $\varphi\phi = 1_{\mathbb{N}_0}$, and similarly proving that $\phi\varphi = 1_{\mathbb{N}_0'}$. Induction on n also shows that

$$\phi(m + n) = \phi(m) + \phi(n)$$

$$\phi(mn) = \phi(m)\phi(n)$$

and

$$m \geq n \Rightarrow \phi(m) \geq \phi(n).$$

Thus the bijection ϕ between $\mathbb{N}_0$ and $\mathbb{N}_0'$ *preserves* the arithmetic and order: we can use it to 'translate' valid results in one into valid results in the other. Such a bijection is called an *order isomorphism*.† In this sense there is only one possible structure for a system satisfying (N1)–(N3): all such systems are order isomorphic. The whole ethos of the natural numbers is encapsulated in three simple axioms!

Now one system which should satisfy these axioms is our intuitive concept $\mathscr{N} \cup \{0\}$, and so this must correspond in the obvious way to $\mathbb{N}_0$. The vital difference is that the properties we expect of $\mathscr{N} \cup \{0\}$ have been built up by example and experience, whereas those of $\mathbb{N}_0$ have been deduced logically from the axioms. Thus all of the usual properties that we expect of $\mathscr{N} \cup \{0\}$ can be given a rigorous justification in $\mathbb{N}_0$. We could, for example, name the elements of $\mathbb{N}_0$ using decimal notation and calculate addition and multiplication tables. At this stage it is more profitable to omit such technicalities on the understanding that they are mere routine.

Counting

As in real life, we can 'count' using the natural numbers. Let

$$\mathbb{N}(n) = \{m \in \mathbb{N} \mid 1 \leqslant m \leqslant n\}$$

for $n \in \mathbb{N}$, and let

$$\mathbb{N}(0) = \varnothing.$$

A set X is said to have n elements ($n \in \mathbb{N}_0$) if there is a bijection

$$f: \mathbb{N}(n) \to X.$$

This models the primitive idea of counting:

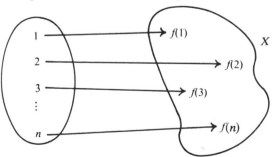

---

† The word 'isomorphism' alone is normally used for a bijection which preserves all relevant arithmetical (algebraic) operations. The word 'order' is used to emphasize that the ordering is also preserved. This usage extends to a variety of mathematical systems.

If we point out the elements $f(1), f(2), \ldots, f(n)$ in turn and call out '1, 2, ..., n' then this is precisely the way in which we count. The useful notational device $\mathbb{N}(0) = \varnothing$ allows us to apply the process to the empty set as well. If a set has n elements for some $n \in \mathbb{N}_0$ then it is said to be *finite*; otherwise it is *infinite*.

This manner of counting does not depend on the order in which we count: given a bijection $f: \mathbb{N}(n) \to X$ and a bijection $g: \mathbb{N}(m) \to X$, we always have $m = n$. To see this, let $\varphi = f^{-1}g$. Then $\varphi: \mathbb{N}(m) \to \mathbb{N}(n)$ is a bijection. We prove by induction that if there is a bijection between $\mathbb{N}(n)$ and $\mathbb{N}(m)$, then $m = n$.

This is certainly true for $m = 0$. Suppose it true for some $m \in \mathbb{N}_0$, and consider a bijection

$$\theta: \mathbb{N}(m+1) \to \mathbb{N}(k).$$

Now $k \neq 0$, or else $m + 1 = 0$ which contradicts (N1). Hence $k = n + 1$ for some $n \in \mathbb{N}_0$. We now construct a bijection $\theta^*: \mathbb{N}(m+1) \to \mathbb{N}(n+1)$ for which $\theta^*(m+1) = n+1$. If it is already the case that $\theta(m+1) = n+1$ then we take $\theta^* = \theta$. If not then $\theta(q) = n+1$ for some $q \leqslant n$, and we define

$$\theta^*(q) = \theta(m+1)$$

$$\theta^*(m+1) = n+1$$

$$\theta^*(r) = \theta(r) \quad \text{otherwise.}$$

Now restrict θ^* to a map

$$\theta^*|_{\mathbb{N}(m)}: \mathbb{N}(m) \to \mathbb{N}(n).$$

This is clearly a bijection, so by induction $m = n$. Hence $m + 1 = n + 1 = k$, completing the induction step.

This validates the intuitive idea of counting within the formal system.

von Neumann's brainwave

As a diversion we may now mention John von Neumann's brilliant method of describing natural numbers, announced in 1923. It is particularly suitable for counting, the number n being defined as a *set* with n elements. To start, there is only one choice for a set with 0 elements, so we put

$$0_v = \varnothing.$$

(Here the suffix v stands for von Neumann.) We now have one object, namely 0_v, so we define

$$1_v = \{0_v\},$$

manifestly a set with 1 element. Now we have two objects 0_v and 1_v, so we define

$$2_v = \{0_v, 1_v\}.$$

It is now clear how to continue. Note that

$$\{0_v, 1_v\} = \{0_v\} \cup \{1_v\} = 1_v \cup \{1_v\}.$$

Having described

$$n_v = \{0_v, 1_v, \ldots, (n-1)_v\}$$

we define

$$(n+1)_v = n_v \cup \{n_v\}$$
$$= \{0_v, \ldots, (n-1)_v\} \cup \{n_v\}$$
$$= \{0_v, \ldots, n_v\}.$$

This can be made more formal as follows. For any set X we let

$$\sigma(X) = X \cup \{X\}$$

be the *successor* of X. This has the bizarre property

$$X \in \sigma(X) \quad \text{and} \quad X \subseteq \sigma(X).$$

Now a set Ω whose elements are sets is called *inductive* if

$$\varnothing \in \Omega$$
$$X \in \Omega \Rightarrow \sigma(X) \in \Omega.$$

To avoid an 'and so on . . . ' definition, von Neumann postulated:

THE AXIOM OF INFINITY.   There exists an inductive set Ω.

This set Ω may be bigger than we require. But if we let $\mathbb{N}_v$ be the intersection of all inductive subsets of Ω, then it is the smallest inductive subset. Hence if $S \subseteq \mathbb{N}_v$ and S is inductive, it follows that $S = \mathbb{N}_v$.

Since $\mathbb{N}_v$ is inductive, we have $\varnothing \in \mathbb{N}_v$, and $X \in \mathbb{N}_v \Rightarrow \sigma(X) \in \mathbb{N}_v$, so $\sigma : \mathbb{N}_v \to \mathbb{N}_v$ is a function. Also $\varnothing \neq \sigma(n)$ for any $n \in \mathbb{N}_v$, since $n \in \sigma(n)$. We shall prove σ is injective.

First note that if $m, n \in \mathbb{N}_v$ and $m \in n$, then $m \subseteq n$. For let

$$S = \{n \in \mathbb{N}_v \mid m \in n \Rightarrow m \subseteq n\}.$$

Trivially $\varnothing \in S$. Suppose $n \in S$ and $m \in \sigma(n)$. Then either $m \in n$ or $m = n$. In either case $m \subseteq n \cup \{n\} = \sigma(n)$. Hence S is an inductive subset of $\mathbb{N}_v$, so $S = \mathbb{N}_v$.

Now suppose that $\sigma(m) = \sigma(n)$. Then $m \cup \{m\} = n \cup \{n\}$. Thus $m \in n \cup \{n\}$ and either $m \in n$ or $m = n$. By the above remark, $m \subseteq n$. Similarly $n \subseteq m$, hence $m = n$ and σ is injective.

Gathering together these remarks, we find that $\mathbb{N}_v$ is a set, $\sigma : \mathbb{N}_v \to \mathbb{N}_v$ is a function, $\varnothing \in \mathbb{N}_v$, and

  (i) $\varnothing \neq \sigma(n)$ for any $n \in \mathbb{N}_v$,

  (ii) $\sigma(m) = \sigma(n) \Rightarrow m = n$,

  (iii) If $S \subseteq \mathbb{N}_v$, $\varnothing \in S$, and $n \in S \Rightarrow \sigma(n) \in S$, then $S = \mathbb{N}_v$.

These are the same as the Peano axioms, with $\mathbb{N}_v$ in place of $\mathbb{N}$, σ in place of s, and $\varnothing$ in place of 0. So von Neumann's idea gives an alternative foundation for the natural numbers, and his axiom of infinity acts as a substitute for the existence axiom for the natural numbers. We could have used this approach instead. However, the simplest way to count in von Neumann's system is to say that a set X has n elements if there is a bijection $f : n_v \to X$, that is,

$$f : \{0_v, 1_v, \ldots, (n-1)_v\} \to X.$$

This corresponds to counting '$0_v, 1_v, \ldots, (n-1)_v$' rather than the more primitive '$1, 2, 3, \ldots, n$' to which we are accustomed.

Other forms of induction

Sometimes the induction step in a proof by induction needs more than the assumption that $P(n)$ be true in order to deduce $P(n+1)$. For example, we may need to know the truth of $P(1), P(2), \ldots, P(n)$ before being able to pass to $P(n+1)$. This situation is governed by the so-called 'General Principle of Induction'. If

  (GP1) $P(0)$ is true,

  (GP2) The truth of $P(m)$ for all $m \in \mathbb{N}_0$ with $m \leq n$ implies the truth of $P(n+1)$,

then $P(n)$ is true for all $n \in \mathbb{N}_0$.

At first sight this seems a genuine extension of the induction principle because the second statement seems more subtle. But if we let $Q(n)$ be the predicate

$$P(0) \,\&\, P(1) \,\&\, \ldots \,\&\, P(n),$$

or more formally,

'for all $m \in \mathbb{N}_0$, $m \leq n$, $P(m)$ is true',

then we find that (GP1) and (GP2) become

(i) $Q(0)$ is true,

(ii) The truth of $Q(n)$ implies the truth of $Q(n+1)$.

Thus the disguise of the 'general' principle is exposed: it is just the ordinary principle for $Q(n)$, and in theory it is no more general than the usual principle of induction. In practice, of course, it sometimes leads to simpler proofs. With it we can prove a highly useful variant of the induction principle:

THEOREM 14 (*Well ordering principle*).   Every non-empty subset of $\mathbb{N}_0$ has a least element.†

PROOF.   More formally, we have to show that if $\varnothing \neq S \subseteq \mathbb{N}_0$, then there exists $a \in S$ such that for all $s \in S$ we have $a \leq s$. For a contradiction, suppose no such a exists. Let $P(n)$ be the predicate $n \notin S$. Then $P(0)$ is true, for $0 \in S$ would imply that 0 is the least element of S by lemma 9. Now suppose that $P(m)$ is true for all $m \leq n$, so that if $m \leq n$ then $m \notin S$. If $s \in S$ then $s > n$, so $s \geq n+1$ by proposition 11. We could not have $n+1 \in S$ since it would then be a least element, so $n+1 \notin S$ and $P(n+1)$ is true. By the general principle of induction $P(n)$ is true for all n, that is, S is empty. This is a contradiction.   □

Another variation of the induction principle starts not at 0 but at some other $k \in \mathbb{N}_0$. If

$P(k)$ is true,

the truth of $P(m)$ for $m \geq k$ implies the truth of $P(m+1)$,

then we may deduce that $P(n)$ is true for all $n \geq k$.

This reduces to the usual principle on putting $Q(n) = P(n+k)$. Most often we meet this with $k=1$. But in the next proposition, which we shall need elsewhere, we require $k=3$.

PROPOSITION 15 (*General associative law*).   If $a_1, \ldots, a_n \in \mathbb{N}_0$, then the sum $a_1 + \ldots + a_n$ takes the same value independently of the manner in which brackets are inserted.

PROOF.   If $n=3$, there are only two methods of bracketing, namely $(a_1 + a_2) + a_3$ and $a_1 + (a_2 + a_3)$. These are equal by theorem 4(a). Suppose the proposition true for some n. Then without ambiguity

---

† That is, an element *belonging to the subset* and $\leq$ every other element of the subset.

we may omit all brackets from a sum of n or fewer numbers. We must therefore consider

$$(a_1 + \ldots + a_k) + (a_{k+1} + \ldots + a_{n+1})$$

and show that the value of this is independent of k. Let

$$a = a_1 + \ldots + a_k$$

$$b = a_{k+1} + \ldots + a_n$$

$$c = a_{n+1}.$$

Then the expression is equal to

$$a + (b + c) = (a + b) + c$$

$$= (a_1 + \ldots + a_n) + a_{n+1}$$

which does not depend on k. This completes the induction step. □

A similar proof works when addition is replaced by multiplication.

Division

Given $m, n \in \mathbb{N}_0$ with $n \neq 0$, it is not always possible to divide n into m and obtain a solution in $\mathbb{N}_0$. For this to happen m must be a multiple of n, that is $m = qn$ for some $q \in \mathbb{N}_0$. If it does not happen, then the division process will yield a remainder.

THEOREM 16 (*Division algorithm*).   Given $m, n \in \mathbb{N}_0$ with $n \neq 0$, there exist unique elements $q, r \in \mathbb{N}_0$ such that $m = qn + r$ and $r < n$.

PROOF.   We use induction on m. Let

$$S = \{m \in \mathbb{N}_0 | m = qn + r \text{ for } q, r \in \mathbb{N}_0, r < n\}.$$

Since $0 = 0n + 0$, we have $0 \in S$. Suppose $m \in S$. Then $m = qn + r$ with $r < n$, and

$$m + 1 = qn + r + 1. \tag{*}$$

Now $r < n$ implies $r + 1 \leq n$. So either $r + 1 = n$, when (*) becomes

$$m + 1 = (q + 1)n + 0$$

or $r + 1 < n$, when (*) becomes

$$m + 1 = qn + (r + 1) \quad \text{with } r + 1 < n.$$

In either case, $m + 1 \in S$, so by induction $S = \mathbb{N}_0$.

To show that q, r are unique suppose that

$$m = qn + r = q'n + r'$$

where $r, r' \in n$. Then

$$qn \leq m < (q+1)n$$

$$q'n \leq m < (q'+1)n.$$

Hence, by transitivity of the order relation, $qn < (q'+1)n$, so by proposition 13 $q < q' + 1$, and then by proposition 11, $q \leq q'$. Similarly $q' \leq q$, so $q = q'$. By proposition 5(a) it now follows that $r = r'$. $\Box$

Factorization

We can now deal with factorization into primes, and in particular prove uniqueness. Only non-zero numbers are of interest, so for the remainder of this chapter we work in $\mathbb{N} = \mathbb{N}_0 \backslash \{0\}$. First some straightforward definitions are required.

We say that $k \in \mathbb{N}$ is a *factor* or *divisor* of $m \in \mathbb{N}$ if there exists $s \in \mathbb{N}$ such that $m = ks$. We write $k|m$. Trivially 1 and m are factors of m; any other factor is called a *proper factor*. We call *m prime* if $m \neq 1$ and m has no proper factors. (We exclude 1 for convenience, for example in the unique factorization theorem which follows.) It is easily seen that a factor k of m must lie in the range $1 \leq k \leq m$, for if $k > m$ then since $s \geq 1$ we find that $ks > m$. A proper factor therefore lies in the range $1 < k < m$.

If k is a factor of two numbers $m, n \in \mathbb{N}$, it is called a *common factor*. Now 1 is always a common factor; if it is the only one we say that m and n are *coprime*. Rather than characterizing the highest common factor as the largest of the common factors (which indeed it is) we choose to define it in a more useful way. We say that $h \in \mathbb{N}$ is the *highest common factor* of $m, n \in \mathbb{N}$ if h is a common factor with the property that any other common factor k must be a factor of h. We write $h = \text{hcf}(m, n)$. The simplest way to prove that any two non-zero natural numbers have a highest common factor is to calculate it explicitly. There is a programme for doing this (called the *Euclidean algorithm* for historical reasons) which depends on the following two facts:

  (i) If $r_1 = q_1 r_2$ then $r_2 = \text{hcf}(r_1, r_2)$,
  (ii) If $r_1 = q_1 r_2 + r_3$ with $r_3 \neq 0$, then $\text{hcf}(r_1, r_2) = \text{hcf}(r_2, r_3)$.

The proofs are easy exercises using the definition of hcf, and in particular (ii) is true since the equation $r_1 = q_1 r_2 + r_3$ shows that any common factor of r_1 and r_2 must also divide r_3, and any common factor of r_2 and r_3 must also divide r_1.

The Euclidean algorithm

To find the hcf of r_1 and r_2 use the division algorithm repeatedly to find q_i, r_i such that

$$r_1 = q_1 r_2 + r_3 \qquad (r_3 < r_2)$$

$$r_2 = q_2 r_3 + r_4 \qquad (r_4 < r_3)$$

$$\cdots$$

$$r_i = q_i r_{i+1} + r_{i+2} \qquad (r_{i+2} < r_{i+1})$$

$$\cdots$$

Since $r_2 > r_3 > r_4 > \ldots$ the process cannot continue indefinitely, for the well ordering principle tells us that the set of numbers concerned has a least element, and it easily follows that at some stage we have $r_{i+2} = 0$, $r_{i+1} \neq 0$. This value of r_{i+1} is a highest common factor for r_1 and r_2. For the statements (i) and (ii) above show that $\mathrm{hcf}(r_1, r_2) = \mathrm{hcf}(r_2, r_3) = \ldots = \mathrm{hcf}(r_i, r_{i+1}) = r_{i+1}$.

As an example we find the hcf of 612 and 221 (allowing the usual operations of arithmetic as part of our contextual technique, since we have seen that they may be formalized within $\mathbb{N}_0$):

$$612 = 2.221 + 170$$
$$221 = 1.170 + 51$$
$$170 = 3.51 + 17$$
$$51 = 3.17$$

Hence $\mathrm{hcf}(612, 221) = 17$.

Notice that this method yields the hcf *without* factorizing the numbers into primes, unlike the method often taught in schools.

PROPOSITION 17.   If h is the hcf of $r_1, r_2 \in \mathbb{N}$, and $n \in \mathbb{N}$, then the hcf of nr_1 and nr_2 is nh.

PROOF.   If we take the steps in the Euclidean algorithm for $\mathrm{hcf}(r_1, r_2)$, as written out above, and multiply through by n, we obtain a system of equations

$$nr_1 = q_1 nr_2 + nr_3 \quad (nr_3 < nr_2)$$

$$nr_1 = q_2 nr_2 + nr_4 \quad (nr_4 < nr_3)$$

$$\cdots$$

$$nr_i = q_i nr_{i+1} \qquad (\text{recalling that } r_{i+2} = 0).$$

The uniqueness of the remainder at each stage implies that this is the Euclidean algorithm for $\text{hcf}(nr_1, nr_2)$, and that the result is

$$nr_{i+1} = n \cdot \text{hcf}(r_1, r_2). \qquad \square$$

From this follows a crucial fact:

LEMMA 18.   If $m, n \in \mathbb{N}$ and p is a prime dividing mn, then either p divides m or p divides n.

PROOF.   Suppose that p does not divide m. Since p is prime, its only factors are $1, p$; so the hcf of p and m must be 1. By proposition 17 the hcf of nm and np is n. But p divides nm and np, so the definition of hcf implies that p divides n. $\square$

COROLLARY 19.   If $m_1, \ldots, m_r \in \mathbb{N}$ and a prime p divides $m_1 \ldots m_r$, then p divides at least one of $m_1, \ldots, m_r$.

PROOF.   Use induction on $r \geqslant 2$. $\square$

The final theorem of this chapter says, in formal terms, that the factorization of a natural number into primes is unique, except for the possibility of writing the factors in a different order.

THEOREM 20 (*Uniqueness of prime factorization*).   Suppose that $m \in \mathbb{N}$, $m \geqslant 2$, and that

$$m = p_1^{e_1} \ldots p_r^{e_r} = q_1^{f_1} \ldots q_s^{f_s}$$

for primes p_i, q_j and natural numbers $e_i, f_j \geqslant 1$. Then $r = s$, and there is a bijection $\phi: \{1, \ldots, r\} \to \{1, \ldots, s\}$ such that for each i, $p_i = q_{\phi(i)}$ and $e_i = f_{\phi(i)}$.

PROOF.   We use induction on $k = e_1 + \ldots + e_r$. If $k = 1$ then $m = p_1$, $r = 1$, $e_1 = 1$. Now p_1 divides the product of q's, hence by corollary 19 divides q_i for some i. Since q_i is prime $p = q_i$. Using 5(b) we may divide through by p_1, obtaining

$$1 = q_1^{f_1} \ldots q_i^{f_i - 1} \ldots q_s^{f_s}$$

which is possible only if $s = 1$, $f_1 = 1$. Hence the two factorizations are given by $m = p_1 = q_1$, and ϕ may be taken to be the identity.

Now suppose the result true for k, and suppose $e_1 + \ldots + e_r = k + 1$. As for $k = 1$, we have $p_1 = q_i$ for some i. It follows that $e_1 = f_i$, or else, dividing out powers of p_1 using proposition 5(b), one side would be divisible by p_1 and the other not. Now we can divide out all powers of

p_1 occurring, to get

$$p_2^{e_2} \ldots p_r^{e_r} = q_1^{f_1} \ldots q_{i-1}^{f_{i-1}} q_{i+1}^{f_{i+1}} \ldots q_s^{f_s}.$$

By induction $r-1 = s-1$, and there is a bijection $\varphi: \{2, \ldots, r\} \to \{1, \ldots, i-1, i+1, \ldots, s\}$ such that $p_j = q_{\varphi(j)}$ and $e_j = f_{\varphi(j)}$ for $j = 2, \ldots, r$. It remains only to define ϕ by

$$\phi(1) = i$$

$$\phi(j) = \varphi(j) \quad \text{for } j = 2, \ldots, r,$$

and the induction step is proved. $\square$

Exercises

The first six exercises are to be considered in the context of this chapter, the remainder apply induction proofs to wider areas of mathematics and should be considered in the appropriate context.

1. Define m^n for $m, n \in \mathbb{N}_0$ by

$$m^0 = 1, \; m^{n+1} = m^n m.$$

Prove by suitable induction arguments that

$$m^{n+r} = m^n m^r$$

$$m^{nr} = (m^n)^r$$

$$(mn)^r = m^r n^r.$$

2. A sequence of natural numbers is a function $s: \mathbb{N} \to \mathbb{N}_0$. We write s_n instead of $s(n)$ and denote s by (s_n). Given a sequence (s_n), the nth *partial sum* σ_n of (s_n) is defined recursively by

$$\sigma_1 = s_1, \qquad \sigma_{n+1} = \sigma_n + s_{n+1}.$$

The sum σ_n is also written as $\sigma_n = s_1 + s_2 + \ldots + s_n$.
Prove by induction that
  (a) $1 + 2 + \ldots + n = \frac{1}{2} n(n+1)$
  (b) $1^2 + 2^2 + \ldots + n^2 = \frac{1}{6} n(n+1)(2n+1)$
  (c) $1^3 + 2^3 + \ldots + n^3 = \frac{1}{4} n^2 (n+1)^2$.

3. Define $n!$ for $n \in \mathbb{N}_0$ by

$$0! = 1, \; (n+1)! = n!(n+1).$$

Prove by induction on n that $(n-r)!r!$ divides $n!$ for all $0 \le r \le n$.

For all $n, r \in \mathbb{N}_0$, $0 \le r \le n$, define $\binom{n}{r} \in \mathbb{N}$ to be $\dfrac{n!}{(n-r)!r!}$.

Show that

$$\binom{n}{0} = 1, \quad \binom{n}{1} = n, \quad \binom{n}{r} = \binom{n}{n-r}$$

and

$$\binom{n}{r} + \binom{n}{r-1} = \binom{n+1}{r}.$$

Use the last equality to prove by induction that for all $a, b, n \in \mathbb{N}_0$:

$$(a+b)^n = a^n + na^{n-1} + \ldots + \binom{n}{r}a^{n-r}b^r + \ldots + \binom{n}{n}b^n.$$

4. Prove by induction, or otherwise,
(a) $1 \cdot 1! + 2 \cdot 2! + \ldots + n \cdot n! = (n+1)! - 1$,

(b) $\binom{n}{0} + \binom{n}{1} + \ldots + \binom{n}{n} = 2^n$,

(c) $\binom{n}{1} + 2\binom{n}{2} + \ldots + n\binom{n}{n} = 2^{n-1}n$.

5. Calculate the highest common factor of 2244 and 2145
(a) by the Euclidean algorithm,
(b) by factorizing 2244 and 2145 into prime factors.

6. The *Fibonacci numbers* (u_n) are defined recursively by

$$u_1 = 1, \quad u_2 = 2, \quad u_{n+1} = u_n + u_{n-1}.$$

Calculate u_3, u_4, u_5, u_6 and u_7. Prove that every natural number is a sum of Fibonacci numbers. Is this expression unique?

7. If $x_1, \ldots, x_n$ are real numbers, prove that

$$|x_1| + \ldots + |x_n| \geq |x_1 + \ldots + x_n|.$$

8. Let p/q be a fraction in lowest terms such that

$$\frac{1}{n+1} < \frac{p}{q} < \frac{1}{n}$$

for a natural number n. Show that $\dfrac{p}{q} - \dfrac{1}{n+1}$ is a fraction which, in its lowest terms, has numerator less than p. Hence, by induction, prove that every proper fraction p/q, where $p < q$, can be written as a finite sum of distinct reciprocals.

$$\frac{p}{q} = \frac{1}{n_1} + \ldots + \frac{1}{n_k}$$

where $n_1, \ldots, n_k$ are natural numbers.
For example $\frac{19}{15} = \frac{1}{2} + \frac{1}{3} + \frac{1}{4} + \frac{1}{6} + \frac{1}{60}$.
Use the technique developed in this question to express $\frac{5}{7}$ as a sum of reciprocals.

9. State and prove analogues of the division algorithm and the Euclidean algorithm for polynomials

$$p(x) = a_n x^n + a_{n-1} x^{n-1} + \ldots + a_0$$

with real coefficients. (Hint: If $a_n \neq 0$, then the degree of $p(x)$ is an element of $\mathbb{N}_0$.)

10. *The Tower of Hanoi* is a puzzle consisting of n discs, of different sizes, which can be placed in three heaps A, B, C. A disc may be 'legally' moved from the top of one pile to the top of another provided that it is not placed on top of a smaller disc. Initially all the discs are placed in one pile A, with the largest at the bottom and in decreasing order of size up the pile; the other two piles are empty. Prove that there exists a sequence of legal moves which will transfer all of the discs to pile B.

11. Are the following valid induction proofs?
   (a) Everybody is bald.
   Proof.   By induction on the number n of hairs. A man with no hairs is clearly bald. Adding one hair to a bald man is not enough to make him not bald, so if a man with n hairs is bald, so is a man with $n + 1$ hairs. By induction, however many hairs a man has, he is bald.
   (b) Everybody has the same number of hairs.
   Proof.   By induction on the number of people. If this is 0 or 1, the statement is clearly true. Assume it for n. Take $n + 1$ people, remove one, then by the induction hypothesis the remaining n people have the same number of hairs. Remove a different one: the remaining n people again have the same number of hairs, so the first one removed has the same number of hairs as the rest. Hence all $n + 1$ people have the same number of hairs.
   (c) If n straight lines are drawn across a circular disk, such that no three meet in the same point, then they divide the disc into 2^n parts.
   Proof.   For $n = 1, 2$, the number of parts is 2, 4. Assume the result true for n. Adding another line divides each region it passes through into two, making 2^{n+1} in all. By induction, the statement is proved.
   (d) $n^2 - n + 41$ is a prime number (positive or negative) for every natural number.
   Proof. $1^2 - 1 + 41 = 41, 2^2 - 2 + 41 = 43, 3^2 - 3 + 41 = 47, 4^2 - 4 + 41 = 53, 5^2 - 5 + 41 = 61, 6^2 - 6 + 41 = 71, \ldots$.
   (e) $1 + 3 + 5 + \ldots + (2n - 1) = n^2 + 1$.
   Proof.   If this is true at n, then add $2n + 1$ to each side to get

$$1 + 3 + 5 + \ldots + (2n - 1) + (2n + 1) = n^2 + 1 + (2n + 1)$$
$$= (n + 1)^2 + 1.$$

This is the same formula with n replaced by $n + 1$, so by induction the formula is true for all natural numbers.

   (f) $2 + 4 + \ldots + 2n = n(n + 1)$
   Proof.   If $2 + 4 + \ldots + 2n = n(n + 1)$, then
$$2 + 4 + \ldots + 2n + 2(n + 1) = n(n + 1) + 2(n + 1),$$
so
$$2 + 4 + \ldots + 2(n + 1) = (n + 1)(n + 2).$$

By induction the formula is true for all n.

12. *Induction with a difference.* The arithmetic mean of the n real numbers $a_1, \ldots, a_n$ is $(a_1 + \ldots + a_n)/n$ and the geometric mean (if they are all non-negative) is $\sqrt[n]{(a_1 a_2 \ldots a_n)}$. Prove that if $a_1, a_2, \ldots, a_n \geqslant 0$, then

$$(a_1 + a_2 + \ldots + a_n)/n \geqslant \sqrt[n]{(a_1 a_2 \ldots a_n)}.$$

You may find a direct induction proof does not work. Try the approach of Cauchy: Let $P(n)$ be the statement '$(a_1 + \ldots + a_n)/n \geqslant \sqrt[n]{(a_1 \ldots a_n)}$ for all real numbers $a_1, \ldots, a_n \geqslant 0$'.

First establish by a straight induction that $0 \leqslant a \leqslant b \Rightarrow 0 \leqslant a^n \leqslant b^n$, and deduce that for $a, b \geqslant 0$, $a^n \leqslant b^n \Rightarrow a \leqslant b$. If $\sqrt[n]{x}$ denotes the positive nth root of $x \geqslant 0$, deduce that for $x, y \geqslant 0$, $x \geqslant y \Leftrightarrow \sqrt[n]{x} \geqslant \sqrt[n]{y}$.

$P(1)$ is trivial, and $P(2)$ may be established by considering the sign of $\frac{1}{4}(a_1 + a_2)^2 - a_1 a_2$. Now prove $P(n) \Rightarrow P(2n)$. (Hint: Use $P(n)$ for $a_1, \ldots, a_n$ and also for $a_{n+1}, \ldots, a_{2n}$ and fit them together using $P(2)$.) Then prove $P(n) \Rightarrow P(n-1)$. (Given $a_1, \ldots, a_{n-1}$, let $a_n = (a_1 + \ldots + a_{n-1})/(n-1)$ and use $P(n)$ to show that

$$\sqrt[n]{(a_1 \ldots a_{n-1} a_n)} \leqslant a_n.$$

Raise to the nth power and simplify to get $P(n-1)$.)
Now deduce that $P(n)$ is true for all $n \in \mathbb{N}$.

13. Proving a statement $P(n)$ true for all $n \in \mathbb{N}$ cannot always be achieved by a simple induction argument. For example, Goldbach's Conjecture that every even integer is the sum of two primes, $2 = 1 + 1$, $4 = 2 + 2$, $6 = 3 + 3$, $8 = 5 + 3$, $10 = 7 + 3, \ldots$ seems plausible (provided that 1 is considered as a prime). Verify Goldbach's Conjecture for every even integer $2n \leqslant 50$. Can you see any pattern that might be amenable for an induction proof? It is not known whether the conjecture is true or false. If you can prove it one way or the other, send us a telegram.

9

The real numbers

FROM our intuitive model $\mathscr{R}$ we can see the kinds of property that would be desirable in a formal model of the real numbers. There would be two binary operations (addition and multiplication) with suitable arithmetical properties, sufficient to allow us to define subtraction and division. Then there would be an order relation, appropriately related to addition and multiplication, and suitably tailored to take account of the presence of negative elements. The final, essential ingredient, is *completeness*, discussed informally in chapter 2. These three facets, arithmetic, order, completeness, suitably expressed, can specify the real numbers completely, just as (N1)–(N3) specify $\mathbb{N}_0$. There are several ways of expressing the required properties, but the experience of the past century's mathematics is that the following system of axioms is one of the best.

Let $\mathbb{R}$ be a set on which are defined two binary operations $+$ and $.$ (called addition and multiplication). If $a, b \in \mathbb{R}$ we call $a + b$ the *sum* of a and b, and $a . b$ the *product*. For traditional reasons we henceforth omit the dot and write ab for the product.

(a) *Arithmetic*

A set $\mathbb{R}$ with binary operations $+$ and $.$ is said to be a *field* if for all $a, b, c \in \mathbb{R}$

(A1) $a + b = b + a$

(A2) $a + (b + c) = (a + b) + c$

(A3) There exists $0 \in \mathbb{R}$ such that $a + 0 = a$ for all $a \in \mathbb{R}$

(A4) Given $a \in \mathbb{R}$ there exists $-a \in \mathbb{R}$ such that $a + (-a) = 0$

(M1) $ab = ba$

(M2) $a(bc) = (ab)c$

(M3) There exists $1 \in \mathbb{R}$, such that $1 \neq 0$, and $1a = a$ for all $a \in \mathbb{R}$

(M4) Given $a \in \mathbb{R}$, $a \neq 0$, there exists $a^{-1} \in \mathbb{R}$ such that $aa^{-1} = 1$

(D) $a(b + c) = ab + ac$.

The elements 0 and 1 are called the *zero* and *unity* elements of $\mathbb{R}$. By virtue of (A1) and (M1) we also have $0 + a = a$, $(-a) + a = 0$, $a1 = a$, $a^{-1}a = 1$, and $(a + b)c = ac + bc$.

We define substraction by

$$a - b = a + (-b)$$

and division by

$$a/b = ab^{-1} \text{ provided } b \neq 0.$$

(b) *Order*

A field $\mathbb{R}$ is said to be *ordered* is there exists a subset $\mathbb{R}^+ \subseteq \mathbb{R}$ such that

(O1) $a, b \in \mathbb{R}^+ \Rightarrow a + b, ab \in \mathbb{R}^+$

(O2) $a \in \mathbb{R} \Rightarrow a \in \mathbb{R}^+$ or $-a \in \mathbb{R}^+$

(O3) $(a \in \mathbb{R}^+) \,\&\, (-a \in \mathbb{R}^+) \Rightarrow a = 0$.

These axioms are designed to relate the order to the arithmetic in the correct way. The set $\mathbb{R}^+$ corresponds to our intuitive idea of the subset of positive elements. The usual order relation is then defined by

$$a \geq b \Leftrightarrow a - b \in \mathbb{R}^+.$$

We shall check later on that this really *is* an order relation.

(c) *Completeness*

An element a of $\mathbb{R}$ is an *upper bound* for a subset $S \subseteq \mathbb{R}$ if $a \geq s$ for all $s \in S$. A set S with an upper bound is said to be *bounded above*. An element λ of $\mathbb{R}$ is a *least upper bound* (lub) for S if

(i) $\lambda \geq s$ for all $s \in S$ (λ is an upper bound)

(ii) $a \geq s$ for all $s \in S \Rightarrow a \geq \lambda$ (λ is the least among the upper bounds).

The final axiom is

(C) If S is a non-empty subset of $\mathbb{R}$ and S is bounded above, then S has a least upper bound in $\mathbb{R}$.

This is the *completeness axiom*. A structure satisfying all thirteen of the above axioms, (A1)–(A4), (M1)–(M4), (D), (O1)–(O3), (C) is called a *complete ordered field*.

If we wish to base our mathematics on the existence axiom for $\mathbb{N}_0$, we must solve the problem of constructing a complete ordered field $\mathbb{R}$ starting from $\mathbb{N}_0$. This can be done, following a similar pattern to the intuitive development used at school level. First the integers $\mathbb{Z}$ are constructed from $\mathbb{N}_0$, and the rationals $\mathbb{Q}$ from $\mathbb{Z}$. This can be done in a very simple manner by using set theory. To construct $\mathbb{R}$ from $\mathbb{Q}$ is a more taxing operation, and the verification of the thirteen axioms is in some places quite difficult.

We shall perform this sequence of constructions now, because it is a part of our mathematical heritage which every mathematician should see once in his life. In retrospect, however, it will appear that the importance of these constructions is to show that *the existence of* N_0, *plus set theory, implies the existence of* $\mathbb{R}$. For practical purposes it is better to take the existence of $\mathbb{R}$ as our basic axiom, and to forget the construction (while remembering that it is possible). The construction itself is a hangover from the nineteenth century when the natural numbers were accepted as the basis of mathematics but the real numbers were imperfectly understood. It was important in that age to show that the real numbers were bona fide mathematical objects, and the demonstration was effected by the construction of $\mathbb{R}$ from N_0. But nowadays, having seen that this can be done, the psychological and philosophical problems involved seem less serious: we assume no more and no less if we postulate that $\mathbb{R}$, rather than N_0, exists. But $\mathbb{R}$ is much more convenient, for it is relatively easy to locate within it a chain of subsets

$$\mathbb{R} \supseteq \mathbb{Q} \supseteq \mathbb{Z} \supseteq N_0$$

giving the rationals, integers, and natural numbers. To do this we have to specify suitable properties to characterize $\mathbb{Z}$ and $\mathbb{Q}$ (N_0 having been taken care of with axioms (N1)–(N3)). Chapter 10 explores this approach to the inner constituents of $\mathbb{R}$, and should make its advantages clear. But now we turn back to the construction of $\mathbb{R}$ from N_0, beginning by putting some flesh on the axioms for arithmetic and order.

Preliminary arithmetical deductions

From our intuitive model $\mathscr{Z}$ of the integers we do not expect all the properties of a field: specifically, not all elements of $\mathbb{Z}$ have multiplicative inverses in $\mathbb{Z}$. But the other arithmetical axioms, namely all except (M4), ought to hold. A set R having two binary operations satisfying (A1)–(A4), (M1)–(M3), and D is called a *ring*†. If further there exists a subset R^+ satisfying (O1)–(O3) we have an *ordered ring*.

We shall now make some elementary deductions from these axioms which, as well as being useful results in their own right, are good practice in the axiomatic style.

---

† More accurately a *commutative* ring. The word 'ring' is usually applied to any system satisfying a less restrictive set of axioms omitting (M1). Since we shall have no cause to deal with non-commutative rings we shall follow current practice and omit the qualifying adjective.

PROPOSITION 1.   If R is a ring, and for some $x \in R$, $a + x = a$ for all $a \in R$, then $x = 0$. If $xa = a$ for all $a \in R$, then $x = 1$.

PROOF.   Put $a = 0$, so that $0 + x = 0$. But $0 + x = x$ by (A3) and (A1), so $x = 0$. Similarly $x = x1 = 1$.   □

This proposition shows that the zero and unity elements of R are unique: no other elements have similar properties. In the same way the negative $-a$ of an element a is uniquely determined:

PROPOSITION 2.   If $x + a = 0$ for elements x, a of a ring R, then $x = -a$.

PROOF.

$$
\begin{aligned}
x &= x + 0 && \text{by (A3)} \\
  &= x + (a + (-a)) && \text{by (A4)} \\
  &= (x + a) + (-a) && \text{by (A2)} \\
  &= 0 + (-a) && \text{since } x + a = 0 \\
  &= -a && \text{by (A4).} \quad \square
\end{aligned}
$$

If R is a field then multiplicative inverses are uniquely determined (for non-zero elements) and the proof is analogous.

PROPOSITION 3.   If R is a ring, then for all $a \in R$, $-(-a) = a$.

PROOF.   By definition $a + (-a) = 0$. By proposition 2, $a = -(-a)$.   □

PROPOSITION 4.   If R is a ring then $a0 = 0a = 0$ for all $a \in R$.

PROOF.

$$
\begin{aligned}
a0 &= a(0 + 0) && \text{by (A3)} \\
   &= a0 + a0 && \text{by (D).}
\end{aligned}
$$

Adding $-(a0)$ to each side, we obtain

$$
\begin{aligned}
0 = a0 + (-a0) &= (a0 + a0) + (-a0) \\
   &= a0 + (a0 + (-a0)) && \text{by (A1)} \\
   &= a0 + 0 && \text{by (A4)} \\
   &= a0 && \text{by (A3).}
\end{aligned}
$$

Then $0a = 0$ by (M1).   □

PROPOSITION 5.   If R is a ring and $a, b \in R$ then $-(ab) = (-a)b = a(-b)$.

PROOF.

$$ab + (-a)b = (a + (-a))b \quad \text{by (D) and (M1)}$$
$$= 0b \qquad\qquad \text{by (A4)}$$
$$= 0 \qquad\qquad \text{by proposition 4.}$$

Hence $(-a)b = -(ab)$ by proposition 2. The rest follows by (M1). $\quad\square$

From here it is easy to make further deductions, such as $(-a)(-b) = ab$, or $(-1)a = -a$. If R is a field we may also prove that $(a^{-1})^{-1} = a$ when $a \neq 0$. Defining subtraction and division as indicated above, we may also verify the expected properties, for example

$$(-a)/b = a/(-b) = -(a/b)$$
$$(a/b) + (c/d) = (ad + bc)/(bd)$$
$$(a/b)(c/d) = (ac)/(bd).$$

The details are left as exercises.

Next we look at order properties. The arithmetic of rings and fields will be considered sufficiently well established that we do not need, any longer, to give chapter and verse when we use it. But for the moment we make explicit mention of the order axioms as we use them.

Preliminary deductions about order

For this section R will be any ordered ring. Its order relation is defined by

$$a \geqslant b \Leftrightarrow a - b \in R^+ \tag{1}$$

as already remarked. It follows at once that $a \geqslant 0 \Leftrightarrow a \in R^+$, so

$$R^+ = \{a \in R \mid a \geqslant 0\}. \tag{2}$$

Using (O1)–(O3) we now establish:

PROPOSITION 6.   The relation $\geqslant$ is a weak order on R.

PROOF.   We must verify the three properties
(WO1) $a \geqslant b \ \& \ b \geqslant c \Rightarrow a \geqslant c$,
(WO2) Either $a \geqslant b$ or $b \geqslant a$,
(WO3) $a \geqslant b \ \& \ b \geqslant a \Rightarrow a = b$.

For (WO1), $a \geqslant b \ \& \ b \geqslant c \Rightarrow a - b, \ b - c \in R^+$. By (O1) $(a - b) + (b - c) \in R^+$, so $a - c \in R^+$, so $a \geqslant c$. (This is our first taste of 'arithmetic without tears': an axiomatic proof that $(a - b) + (b - c) = a - c$ takes several steps, all omitted here.)

For (WO2), (O2) implies that either $a - b \in R^+$ or $b - a = -(a - b) \in R^+$. Therefore $a \geqslant b$ or $b \geqslant a$.

For (WO3), if both $a - b$ and $b - a \in R^+$ then $a - b = 0$ by (O3), so $a = b$. $\square$

The order relation behaves appropriately with respect to the arithmetic:

PROPOSITION 7.   For all $a, b, c, d \in R$,
(a) $a \geqslant b \ \& \ c \geqslant d \Rightarrow a + c \geqslant b + d$,
(b) $a \geqslant b \geqslant 0 \ \& \ c \geqslant d \geqslant 0 \Rightarrow ac \geqslant bd$.

PROOF.   Translate the definition of $\geqslant$ by (1) and use arithmetic. $\square$

In the definition of an ordered field (or ordered ring) we can replace (O1)–(O3) by the properties stated in propositions 6 and 7, and use the relation $\geqslant$ to define the set R^+ by working (2) the other way round. Which we use is a matter of taste.

The modulus can be defined in an ordered ring by setting

$$|a| = \begin{cases} a \text{ if } a \in R^+ \\ -a \text{ if } -a \in R^+. \end{cases}$$

It can then be proved that $|a| \geqslant 0$ for all $a \in R$, and by repeating the argument of chapter 2 in this formal context, that

$$|a + b| \leqslant |a| + |b|$$
$$|ab| = |a| \, |b|.$$

Now we have enough technique to carry out the construction of the integers, rationals, and reals.

The construction of the integers

To get from $\mathbb{N}_0$ to integers we must introduce some negative elements. In fact what we do is consider differences $m - n$ of natural numbers. These are definable *as natural numbers* provided $m \geqslant n$, and our task is to give them a meaning if $n > m$. If $m, n, r, s \in \mathbb{N}_0$ and $m \geqslant n, r \geqslant s$, then

$$m - n = r - s \Leftrightarrow m + s = r + n.$$

Now the right-hand side makes sense *without* restrictions on m, n, r, s. This gives a clue. We take the set $\mathbb{N}_0 \times \mathbb{N}_0$ of ordered pairs (m, n) and define on it a relation $\sim$ by

$$(m, n) \sim (r, s) \Leftrightarrow m + s = r + n.$$

To verify that this is an *equivalence* relation requires only arithmetic in $\mathbb{N}_0$. Having done this we define the integers $\mathbb{Z}$ to be the set of equivalence classes. The *idea* is that the equivalence class of (m, n) should correspond to our intuitive $m - n$, and the formal working out of the idea runs as follows. Let $\langle m, n \rangle$ denote the equivalence class of (m, n). Then $\langle m, n \rangle = \langle r, s \rangle \Leftrightarrow m + s = r + n$. We define addition and multiplication on $\mathbb{Z}$ by†

$$\langle m, n \rangle + \langle p, q \rangle = \langle m + p, n + q \rangle$$
$$\langle m, n \rangle \langle p, q \rangle = \langle mp + nq, mq + np \rangle.$$

The problem is to check that these are well defined in the sense of chapter 4. So suppose that $\langle m, n \rangle = \langle m', n' \rangle$ and $\langle p, q \rangle = \langle p', q' \rangle$. Then $m + n' = m' + n, p + q' = p' + q$. Now

$$(m + p) + (n' + q') = (m + n') + (p + q')$$
$$= (m' + n) + (p' + q)$$
$$= (m' + p') + (n + q).$$

Hence $\langle m + p, n + q \rangle = \langle m' + p', n' + q' \rangle$, and addition is well defined. Multiplication is treated in the same way.

It is now a simple but longwinded exercise to show that $\mathbb{Z}$ is an ordered ring, taking

$$\mathbb{Z}^+ = \{\langle m, n \rangle \in \mathbb{Z} \mid m \geqslant n \text{ in } \mathbb{N}_0\}.$$

PROPOSITION 8.   With the above operations, $\mathbb{Z}$ is an ordered ring.

PROOF.   We must check the axioms (A1)–(A4), (M1)–(M3), (D), and (O1)–(O3). In all cases we use the definition of $\mathbb{Z}$ to restate the required property in $\mathbb{N}_0$, and verify it by arithmetic. As a token we prove (A1).

---

† Think of $\langle m, n \rangle$ as '$m - n$'. Then the definitions are found by translating

$$(m - n) + (p - q) = (m + p) - (n + q)$$
$$(m - n)(p - q) = (mp + nq) - (mq + np).$$

Let $a = \langle m, n \rangle$, $b = \langle p, q \rangle$. Then

$$a + b = \langle m + p, n + q \rangle$$
$$= \langle p + m, q + n \rangle \quad \text{by arithmetic in } \mathbb{N}_0$$
$$= b + a.$$

The remaining axioms are sometimes more awkward, but follow similar lines: the reader should get out pencil and paper and check them! Mathematics is not a spectator sport. $\square$

Our next step is to recover the usual notation of integers as positive or negative natural numbers. Any element of $\mathbb{Z}^+$ is of the form $\langle m, n \rangle$ where $m \geq n$, so can be written as $\langle m - n, 0 \rangle$. Thus every element of $\mathbb{Z}^+$ is of the form $\langle r, 0 \rangle$ for $r \in \mathbb{N}_0$. Now axiom (O2) tells us that for any $a \in \mathbb{Z}$, either $a \in \mathbb{Z}^+$ or $-a \in \mathbb{Z}^+$, hence either $a = \langle r, 0 \rangle$ or $a = -\langle r, 0 \rangle = \langle 0, r \rangle$.

Define a map $f : \mathbb{N}_0 \to \mathbb{Z}^+$ by $f(n) = \langle n, 0 \rangle$. It is easily seen that f is a bijection, and that

$$f(m + n) = f(m) + f(n)$$
$$f(mn) = f(m)f(n)$$
$$m \geq n \Leftrightarrow f(m) \geq f(n),$$

that is, f is an order isomorphism, in the sense of the previous chapter.

This leads to a technical problem. We do not have $\mathbb{N}_0 \subseteq \mathbb{Z}$, as we might have hoped: instead $\mathbb{N}_0$ is order isomorphic to $\mathbb{Z}^+ \subseteq \mathbb{Z}$. What this means is that although $\mathbb{N}_0$ and $\mathbb{Z}^+$ are definitely *different* mathematical objects (the latter is a set of equivalence classes of ordered pairs) the distinction between them becomes irrelevant as soon as we pass beyond the contructional stage: they have exactly the same mathematical structure as regards arithmetic and order.

One way to deal with this is to note that $\mathbb{Z}^+$ satisfies the axioms (N1)–(N3) and so will do just as well as $\mathbb{N}_0$. So having selected a temporary $\mathbb{N}_0$ we construct $\mathbb{Z}$ and forget about the old $\mathbb{N}_0$. The trouble is, we have to do this again with $\mathbb{Q}$ and $\mathbb{R}$, and the whole proccss makes a mountain out of a molehill. Equally, to retain the distinctions between $\mathbb{N}_0$ and $\mathbb{Z}^+$, and similar ones at later stages in the construction, involves littering the landscape with unsightly isomorphisms which confuse what is really a simple situation.

A more satisfactory method is to change the notation to take advantage of the isomorphism. Namely, we allow the notations

$$n \text{ for } \langle n, 0 \rangle$$

$$-n \text{ for } \langle 0, n \rangle,$$

and appeal to the isomorphism to show that this leads to no conflicting results in arithmetic or order. What this amounts to is that we cut out $\mathbb{Z}^+$ and replace it by $\mathbb{N}_0$. We could even do this in a formal set-theoretic way, by defining operations on $(\mathbb{Z} \backslash \mathbb{Z}^+) \cup \mathbb{N}_0$. In the resulting structure, which we rename $\mathbb{Z}$, we then have $\mathbb{N}_0 \subseteq \mathbb{Z}$. (The diagram below should make the idea clear enough.)

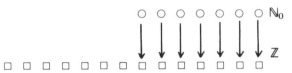

The construction of the rationals

Here the pattern is very similar. Starting now from $\mathbb{Z}$, we need to introduce a larger set $\mathbb{Q}$ on which quotients m/n are defined. To do this, let S be the set of all ordered pairs (m, n) where $m, n \in \mathbb{Z}$ and $n \neq 0$. Define a relation $\sim$ by

$$(m, n) \sim (p, q) \Leftrightarrow mq = np.$$

(This is inspired by the desired property that $m/n = p/q$ if and only if $mq = np$.) Now define $\mathbb{Q}$ to be the set of equivalence classes and, anticipating the final result, use the notation m/n for the equivalence class of (m, n). Define operations by

$$m/n + p/q = (mq + np)/nq$$

$$(m/n)(p/q) = mp/nq.$$

THEOREM 9. These operations define on $\mathbb{Q}$ the structure of a field.

PROOF. The details are left to the reader. First it is necessary to check that the operations are well defined. Then a massive list of axioms have to be waded through. It is really only worth doing these for yourself: it is seldom helpful to read through someone else's long calculations when they are routine. But anyway, here's a hint: for $m, n \neq 0$ the multiplicative inverse of m/n is n/m. $\square$

Now we define an ordering, by putting

$$\mathbb{Q}^+ = \{m/n \in \mathbb{Q} | m, n \in \mathbb{Z}^+, n \neq 0\}.$$

THEOREM 10.   With the above definition, $\mathbb{Q}$ is an ordered field.

PROOF.   Once more the authors pass the buck: you do it!   □

Now there is a map $g: \mathbb{Z} \to \mathbb{Q}$, defined by $g(n) = n/1$. Once again,

$$g(m + n) = g(m) + g(n)$$
$$g(mn) = g(m)g(n)$$
$$m \geq n \Rightarrow g(m) \geq g(n)$$

for all $m, n \in \mathbb{Z}$; so g is an order isomorphism again. As for $\mathbb{N}$ and $\mathbb{Z}$, we can now think of $\mathbb{Z}$ as being a subset of $\mathbb{Q}$, even though this is not exactly the case.

Since every rational m/n can be written as $(m/1)(1/n) = (m/1)(n/1)^{-1}$, it follows that identifying n with $n/1$ does not lead to any conflict in notation, and corresponds to the usual intuitive model.

The content of the chapter so far is really just algebraic fun and games. But now things become more serious.

The real numbers

It is possible, though technically awkward, to construct the real numbers as infinite decimals, along the lines of chapter 2. However, we saw there that the use of approximating sequences of rationals has technical advantages. Monotonic sequences are especially easy to handle, but in fact we shall use the more general "Cauchy sequences" to be defined in a moment. As in the previous sections, many of the routine details will be omitted; and the excuse is the same: that by so doing the broad outline becomes more easily visible, and the details merge into the conceptual background.

Sequences of rationals

A *sequence* of rationals may be formally defined as a function

$$s: \mathbb{N} \to \mathbb{Q}.$$

We write s_n for $s(n)$ and denote the sequence by $(s_n)_{n \in \mathbb{N}}$, or by $(s_1, s_2, s_3, \ldots)$, or just by (s_n).

Let $\mathbb{S}$ be the set of all sequences of rationals. We may define addition and multiplication within $\mathbb{S}$ by

$$(a_n) + (b_n) = (a_n + b_n)$$

$$(a_n)(b_n) = (a_n b_n).$$

LEMMA 11.   With these operations, $\mathbb{S}$ is a ring.

PROOF.   The identity is $(1, 1, 1, \ldots)$, the zero $(0, 0, 0, \ldots)$, and the additive inverse of (a_n) is $(-a_n)$. All verifications are absolutely routine. $\square$

Notice, though, that $\mathbb{S}$ is not a field. If all terms s_n are non-zero, then (s_n) has multiplicative inverse $(1/s_n)$. But if any term $s_n = 0$, then no inverse can exist. For example, $(0, 1, 1, \ldots)$ cannot have an inverse $(b_1, b_2, b_3, \ldots)$ since

$$(0, 1, 1, 1, \ldots)(b_1, b_2, b_3, \ldots) = (0, b_2, b_3, \ldots) \neq (1, 1, 1, \ldots).$$

As we saw in chapter 2, every real number may be viewed as the 'limit' of a sequence of rationals. In the present context we can take over the definition of convergence given in that chapter provided we insist that the ε in the definition is rational.

DEFINITION.   A sequence of rationals (s_n) is said to converge to $l \in \mathbb{Q}$ if, given any $\varepsilon \in \mathbb{Q}$, $\varepsilon > 0$, there exists $N \in \mathbb{N}$ such that

$$n > N \Rightarrow |s_n - l| < \varepsilon.$$

This is not totally satisfactory: convergence to a rational limit is not what really interests us. For the sake of argument let us assume that it makes sense to talk of a sequence of rationals converging to a real limit. Certainly this is so in our intuitive models $\mathcal{2} \subseteq \mathcal{R}$. The catch is that *formally* we do not know what the limit is. Nonetheless, if (s_n) were to converge to a real number l, then we would have some N such that

$$|s_n - l| < \varepsilon \quad \text{for all } n > N.$$

Hence also

$$|s_m - l| < \varepsilon \quad \text{for all } m > N.$$

Combining the two inequalities we obtain

$$|s_m - s_n| < 2\varepsilon \quad \text{for all } m, n > N.$$

Now *this* statement does not involve the hypothetical real number l.

We tidy things up by starting again with $\frac{1}{2}\varepsilon$ instead of ε, and thereby obtain the essential idea:

DEFINITION.   A sequence (s_n) of rational numbers is a *Cauchy sequence* if for any rational $\varepsilon > 0$ there exists N such that

$$m, n > N \Rightarrow |s_m - s_n| < \varepsilon.$$

Intuitively the terms of such a sequence get closer and closer together. (Augustin Cauchy was a prolific nineteenth century French mathematician who made extensive use of such sequences, although he was probably not the originator of the notion.)

Cauchy sequences, thought of as sequences of rational approximations to a real number, provide the raw material for a formal construction of real numbers themselves.

LEMMA 12.   Every Cauchy sequence is bounded.

PROOF.   Taking $\varepsilon = 1$, there exists N such that $|s_n - s_m| < 1$ for $m, n > N$. Thus for all $n > N$ we have $|s_n - s_{N+1}| < 1$, that is, $|s_n| < |s_{N+1}| + 1$. Hence for all $n \in \mathbb{N}$,

$$|s_n| \leqslant \max\{|s_1|, |s_2|, \ldots, |s_N|, |s_{N+1}| + 1\}. \quad \square$$

LEMMA 13.   If (a_n) and (b_n) are Cauchy sequences, then so are $(a_n + b_n)$, $(a_n b_n)$, and $(-a_n)$.

PROOF.   If $\varepsilon > 0$ is rational, there exist N_1 and N_2 such that

$$m, n > N_1 \Rightarrow |a_m - a_n| < \tfrac{1}{2}\varepsilon,$$
$$m, n > N_2 \Rightarrow |b_m - b_n| < \tfrac{1}{2}\varepsilon.$$

So for $m, n > N = \max(N_1, N_2)$ we have

$$
\begin{aligned}
|(a_m + b_m) - (a_n + b_n)| &= |(a_m - a_n) + (b_m - b_n)| \\
&\leqslant |a_m - a_n| + |b_m - b_n| \\
&< \tfrac{1}{2}\varepsilon + \tfrac{1}{2}\varepsilon \\
&= \varepsilon,
\end{aligned}
$$

so $(a_n + b_n)$ is Cauchy.

To show that $(a_n b_n)$ is Cauchy note that by lemma 12, there exist A, B such that $|a_n| < A$ and $|b_n| < B$ for all $n \in \mathbb{N}$, with $A, B \in \mathbb{Q}$. Using a little foresight (the authors have seen this proof before!), given $\varepsilon \in \mathbb{Q}$, $\varepsilon > 0$, we note that $\varepsilon/(A + B) \in \mathbb{Q}$, $\varepsilon/(A + B) > 0$; so we can find N_1

and N_2 such that

$$m, n > N_1 \Rightarrow |a_m - a_n| < \varepsilon/(A + B),$$
$$m, n > N_2 \Rightarrow |b_m - b_n| < \varepsilon/(A + B).$$

If $m, n > N = \max(N_1, N_2)$ then both inequalities hold, so

$$|a_m b_m - a_n b_n| = |(a_m - a_n)b_m + a_n(b_m - b_n)|$$
$$\leq |a_m - a_n||b_m| + |a_n||b_m - b_n|$$
$$< (\varepsilon/(A + B))B + A \cdot \varepsilon/(A + B)$$
$$= \varepsilon.$$

So $(a_n b_n)$ is Cauchy.

Finally $(-a_n)$ may be proved Cauchy either by a direct calculation, or by putting $b_n = -1$ for all n in the above. $\square$

Letting $\mathscr{C}$ denote the set of all Cauchy sequences we now have:

PROPOSITION 14. With addition and multiplication of sequences as defined, $\mathscr{C}$ is a ring.

PROOF. If $(a_n), (b_n) \in \mathscr{C}$ then lemma 13 says that $(a_n)+(b_n)$, $(a_n)(b_n)$, and $-(a_n) \in \mathscr{C}$. Clearly the zero sequence $(0, 0, \ldots)$ and unit sequence $(1, 1, \ldots) \in \mathscr{C}$. Looking at the axioms for a ring we see that this takes care of (A3), (A4), and (M3). The remaining axioms hold since, by lemma 11, they hold for all sequences of rationals. $\square$

However, we still do not have a field, for a sequence like $(0, 1, 1, 1, \ldots)$ is Cauchy, non-zero, and has no inverse. To overcome this we take note of another problem: that intuitively speaking, different Cauchy sequences can converge to the same limit. Both difficulties evaporate when we introduce one further concept:

DEFINITION. A sequence (s_n) of rationals is called a *null sequence* if it converges to 0, that is, if for all rational $\varepsilon > 0$ there exists N such that $|s_n| < \varepsilon$ whenever $n > N$.

If two sequences (a_n) and (b_n) tend to the same limit l, then it is easy to see that the sequence $(a_n - b_n)$ is null. This inspires an equivalence relation on $\mathscr{C}$:

$$(a_n) \sim (b_n) \Leftrightarrow (a_n - b_n) \text{ is null.}$$

To check that this *is* an equivalence relation, observe that the facts $(a_n) \sim (a_n)$ and $(a_n) \sim (b_n) \Rightarrow (b_n) \sim (a_n)$ are utterly trivial. If $(a_n) \sim (b_n)$

and $(b_n) \sim (c_n)$ then $(a_n - b_n)$ and $(b_n - c_n)$ are null, that is converge to 0. By the argument of the first theorem of chapter 7, this implies that $((a_n - b_n) + (b_n - c_n))$ converges to zero, that is, $(a_n - c_n)$ is null, so $(a_n) \sim (c_n)$.

Now let $\mathbb{R}$ be the set of equivalence classes of Cauchy sequences, and denote the equivalence class containing (s_n) by $[s_n]$.

The operations of addition and multiplication are transferred to $\mathbb{R}$ by defining

$$[a_n] + [b_n] = [a_n + b_n]$$

$$[a_n][b_n] = [a_n b_n].$$

These operations are well defined. For if $[a_n] = [a'_n]$ and $[b_n] = [b'_n]$ then $(a_n - a'_n)$ and $(b_n - b'_n)$ are null. Hence $((a_n + b_n) - (a'_n + b'_n))$ is null, so $[a_n + b_n] = [a'_n + b'_n]$. Multiplication is a little less straightforward to handle. By lemma 12 there exist rationals A, B such that

$$|a_n| < A, |b'_n| < B, \text{ for all } n \in \mathbb{N}.$$

Now given $\varepsilon > 0$ we can find N_1, N_2 such that

$$n > N_1 \Rightarrow |a_n - a'_n| < \varepsilon/(A + B),$$

$$n > N_2 \Rightarrow |b_n - b'_n| < \varepsilon/(A + B).$$

If $n > N = \max(N_1, N_2)$ then

$$|a_n b_n - a'_n b'_n| = |a_n(b_n - b'_n) + (a_n - a'_n)b'_n|$$

$$\leq |a_n||b_n - b'_n| + |a_n - a'_n||b'_n|$$

$$< A \cdot \varepsilon/(A + B) + \varepsilon/(A + B) \cdot B$$

$$= \varepsilon.$$

Thus $(a_n b_n - a'_n b'_n)$ is null, so $[a_n b_n] = [a'_n b'_n]$.

THEOREM 15.   With these operations, $\mathbb{R}$ is a field.

PROOF.   Verification of axioms (A1)–(A4), (M1)–(M3), and (D) is a mundane task. The zero element is $[0]$, the unity $[1]$, and the negative of $[a_n]$ is $[-a_n]$. The meat of the proof is (M4): if $[a_n] \neq [0]$, then $[a_n]$ has an inverse in $\mathbb{R}$.

Now $[a_n] \neq [0]$ if and only if (a_n) is not null. Thus there exists some $\varepsilon > 0$ for which there does not exist N with the property

$n > N \Rightarrow |a_n| < \varepsilon$. Less tortuously, for this value of ε, for any $N \in \mathbb{N}$ there exists $n \in \mathbb{N}$ with $n > N$ and $|a_n| \geq \varepsilon$.

But (a_n) is Cauchy, so there exists N_1 such that if $m, n > N_1$ then $|a_m - a_n| < \frac{1}{2}\varepsilon$. Let $N_2 = \max(N, N_1)$. Then for any $m > N_2$ we find $n > N_2$ such that $|a_n| \geq \varepsilon$, and then $|a_m - a_n| < \frac{1}{2}\varepsilon$. So $|a_m| > \frac{1}{2}\varepsilon$. (If you don't see why: if $|a_m| \leq \frac{1}{2}\varepsilon$ then we have $|a_n| = |(a_n - a_m) + a_m| \leq |a_n - a_m| + |a_m| < \frac{1}{2}\varepsilon + \frac{1}{2}\varepsilon = \varepsilon$, which is a contradiction.)

In particular, $a_m \neq 0$ for $m > N_2$. Define a sequence (b_n) by

$$b_n = \begin{cases} 0 & \text{if } n \leq N_2 \\ 1/a_n & \text{if } n > N_2, \end{cases}$$

so that

$$a_n b_n = \begin{cases} 0 & \text{if } n \leq N_2 \\ 1 & \text{if } n > N_2. \end{cases}$$

Then $(1 - a_n b_n)$ is a null sequence, actually *equal* to zero when $n > N_2$, so that $[a_n b_n] = [1]$. Thus $[a_n]$ has an inverse, and $\mathbb{R}$ is a field. $\square$

The ordering on $\mathbb{R}$

To define the ordering on $\mathbb{R}$ there is a new problem. A sequence which tends to a positive limit may involve negative terms as well as positive ones. The next lemma permits a suitable, though slightly unnatural, definition which works.

LEMMA 16.   Every non-null Cauchy sequence of rationals (a_n) has the property that there exists $\varepsilon > 0$ and $N \in \mathbb{N}$ such that $|a_n| > \varepsilon$ for all $n > N$.

PROOF.   This is the second and third paragraphs of the proof of theorem 15, with $\frac{1}{2}\varepsilon$ replaced by ε. $\square$

DEFINITION. $[a_n] \in \mathbb{R}^+$ if and only if either
   (i) $[a_n] = [0]$, or
   (ii) There exists $\varepsilon > 0$, $\varepsilon \in \mathbb{Q}$, such that for some $N \in \mathbb{N}$, whenever $n > N$ then $a_n > \varepsilon$.

Thus a strictly positive Cauchy sequence must have all its terms, after some point, not only positive, but a definite amount greater than zero (namely ε).

THEOREM 17. $\mathbb{R}$ is an ordered field.

PROOF.   It is clear that if $[a_n], [b_n] \in \mathbb{R}^+$ then $[a_n + b_n] \in \mathbb{R}^+$, and that $[a_n b_n] \in \mathbb{R}^+$. It is also clear that if $[a_n] \in \mathbb{R}^+$ and $[-a_n] \in \mathbb{R}^+$, then $[a_n] = [0]$.

It remains to show that $[a_n] \in \mathbb{R}^+$ or $[-a_n] \in \mathbb{R}^+$. Either $[a_n] = [0] \in \mathbb{R}^+$, or for some $N \in \mathbb{R}$, $\varepsilon > 0$, we have $|a_n| > \varepsilon$ for $n > N$, by lemma 16. Also, since (a_n) is Cauchy, we have $|a_m - a_n| < \frac{1}{2}\varepsilon$ for m, $n > N_1$. Thus for $n > \max(N, N_1)$ *the sign of a_n cannot change.* Otherwise we should have $a_n > \varepsilon$, $a_{n+1} < -\varepsilon$, and $|a_n - a_{n+1}| > 2\varepsilon$; or else $a_n < -\varepsilon$, $a_{n+1} > \varepsilon$, with the same conclusion.

Thus either there exists $N_2 = \max(N, N_1)$ such that $a_n > \varepsilon$ for all $n > N_2$, in which case $[a_n] \in \mathbb{R}^+$; or else $a_n < -\varepsilon$ for all $n > N_2$, in which case $[-a_n] \in \mathbb{R}^+$. $\square$

The completeness of $\mathbb{R}$

By far the trickiest property to establish is completeness. We begin by noting that if we define

$$\hat{q} = [q, q, q, \ldots] \text{ for } q \in \mathbb{Q}$$

then $\mathbb{Q}$ is order isomorphic to the subset $\{\hat{q} \in \mathbb{R} | q \in \mathbb{Q}\}$ of $\mathbb{R}$, an assertion which is readily checked.

LEMMA 18.   If $[a_n] \in \mathbb{R}$ then $[a_n] < \hat{p}$ for some $p \in \mathbb{Z}$.

PROOF.   By lemma 12, $|a_n| < A$ for some $A \in \mathbb{Q}$. If p is an integer greater than or equal to $A + 1$, then $[a_n] < \hat{p}$. $\square$

THEOREM 19. $\mathbb{R}$ is a complete ordered field.

PROOF.   Only completeness remains to be established. So let $X \neq \varnothing$ be a subset of $\mathbb{R}$, and suppose that X is bounded above by an element $[x_n]$ of $\mathbb{R}$. By lemma 18, $[x_n] < \hat{p}$ for some $p \in \mathbb{Z}$, hence X is bounded above by $\hat{p}$.

For each $m \in \mathbb{N}$ we define a rational number r_m as follows. Let k_m be the smallest integer such that $(\widehat{k_m/2^m})$ is an upper bound for X, and let $r_m = k_m/2^m$. (Here the notation $\widehat{k_m/2^m}$ is just a typographically convenient way of writing $\hat{q}$ where $q = k_m/2^m$.)

First we must justify this procedure. The set S of all integers j_m for which $\widehat{j_m/2^m}$ is an upper bound for X is non-empty, since $j_m = 2^m p$ will do. On the other hand there exists some integer h for which $\widehat{h/2^m}$ is not an upper bound. For if $[a_n] \in X$ then we may find $N \in \mathbb{N}$, $A \in \mathbb{Q}$ such that for all $n > N$,

$$-A < a_n < A.$$

If we make $h/2^m < -A$ then $\widehat{h/2^m}$ is not an upper bound for X. It follows that every $j_m \in S$ is greater than h. Let T be the set of natural numbers $\{j_m - h | j_m \in S\}$: this is non-empty, hence by the well ordering principle has a least element l. Put $k_m = l + h$.

Now we may picture the rational numbers $r_m = k_m/2^m$ as the smallest rational of the form (integer)/2^m for which $\hat{r}_m$ is an upper bound for X.

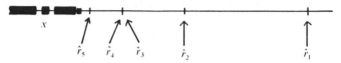

By definition we have

$$2k_m - 2 < k_{m+1} \leqslant 2k_m.$$

For if

$$\overbrace{(2k_m - 2)/2^{m+1}}$$

is an upper bound for X, so is

$$\overbrace{(k_m - 1)/2^m},$$

contrary to k_m being the smallest integer with this property; and on the other hand

$$\overbrace{2k_m/2^{m+1}}$$

is an upper bound, so $k_{m+1} \leqslant 2k_m$. It follows that

$$r_m - \frac{1}{2^{m+1}} \leqslant r_{m+1} \leqslant r_m.$$

But it is an easy consequence of this that (r_n) is a *Cauchy* sequence, for

$$|r_m - r_{m+s}| \leqslant 2^{-m-1} + \ldots + 2^{-m-s} < 2^{-m}.$$

We are almost there, now: put $r = [r_n] \in \mathbb{R}$. Then the picture above leads us to expect that r will be the desired least upper bound.

That r is *an* upper bound follows by a simple argument from the fact that each $\hat{r}_m$ is an upper bound. Assume it is not the least upper bound, so that $t < r$ is an upper bound for x. Then $t = [t_n]$, and by the definition of ordering on $\mathbb{R}$, there exists $\varepsilon > 0$, $\varepsilon \in \mathbb{Q}$, such that $r_n - t_n > \varepsilon$ for $n > N$. But for large enough m we have $1/2^m < \varepsilon$, so we may reduce r_m to $r_m - 1/2^m$ while retaining an upper bound. But this reduces k_m to $k_m - 1$, contradicting the defining property of k_m. Hence r is a least upper bound, and $\mathbb{R}$ is complete. $\square$

As before, we do not actually have $\mathbb{Q} \subseteq \mathbb{R}$. But we do have an order isomorphism which allows us to replace the elements $\hat{q}$ of $\mathbb{R}$ by elements q of $\mathbb{Q}$, thus making $\mathbb{Q}$ a genuine subset of $\mathbb{R}$. So finally we have a chain of number systems

$$\mathbb{N} \subseteq \mathbb{N}_0 \subseteq \mathbb{Z} \subseteq \mathbb{Q} \subseteq \mathbb{R}$$

as intended.

Exercises

1. First some book-keeping.
  (a) Write out a full proof of proposition 6, including the properties of arithmetic assumed in that context.
  (b) Complete the proof of proposition 8.
  (c) Prove theorem 10.

2. If R is a ring, for $x \in R$, $n \in \mathbb{N}_0$ define $\hat{n}, x^n \in R$ recursively by

$$\hat{0} = 0_R, \widehat{n+1} = \hat{n} + 1_R,$$
$$x^0 = 1_R, x^{n+1} = x^n x.$$

If $\binom{n}{r}$ is the binomial coefficient $n!/[r!(n-r)!]$, prove that for all $x, y \in R$

$$(x+y)^n = x^n + \hat{n}x^{n-1}y + \ldots + \widehat{\binom{n}{r}}x^{n-r}y^r + \ldots + y^n.$$

3. If $p \in \mathbb{N}$ is a prime and $\hat{p} = 0_R$ in R, (as is, for instance, the case in $\mathbb{Z}_p$), show that

$$(x+y)^p = x^p + y^p.$$

Give an example of a ring R where $\hat{n} = 0_R$, but $(x+y)^n \neq x^n + y^n$.

4. If R is an ordered ring, show that

$$x^2 - \hat{5}x + \hat{6} \geqslant 0_R$$

if and only if $x \geqslant \hat{3}$ or $x \leqslant \hat{2}$.

5. Use the Euclidean algorithm to prove that if $m, n \in \mathbb{N}$ are coprime, then there exist $a, b \in \mathbb{N}$ such that

$$am + bn = 1.$$

Find a, b when $m = 1008$, $n = 1375$.

6. Prove formally that every positive rational number can be written uniquely in the form m/n for *coprime* $m, n \in \mathbb{N}$. This is called 'expressing a fraction in its lowest terms'. If two rationals p/q, r/s are in lowest terms, is $(ps+qr)/(qs)$? What about $(pr)/(qs)$?

Show that if p/q is in lowest terms, so is p^2/q^2. Use the uniqueness of expression in lowest terms to give a streamlined proof that $\sqrt{2}$ is irrational.

7. Prove the following results in any ordered field (which we suppose contains $\mathbb{Q}$).

(a) $a \leqslant b \Leftrightarrow -b \leqslant -a$,
(b) $a < b \Leftrightarrow -b < -a$,
(c) $-1 < 0 < 1$,
(e) if $a \neq 0$, then $a^2 > 0$,
(f) $0 < a \leqslant b \Rightarrow 0 < b^{-1} \leqslant a^{-1}$,
(g) if $a < 0$ and $b < 0$, then $ab > 0$.

8. Prove that every non-empty *finite* subset X of an ordered field contains a smallest element and a largest element. (A smallest element is an element $x \in X$ such that $x \leqslant y$ for all $y \in X$; a largest element is defined similarly.) Is the same true if we drop the condition that X be finite?

9. In the definition of the order relation on $\mathbb{R}$, why is it not a good idea to define

$$[a_n] \geqslant 0 \text{ if } \exists N \in \mathbb{R}, \text{ such that } a_n \geqslant 0 \text{ for all } n > N?$$

10. Let $a_n = \frac{1}{2} - \frac{1}{6} + \ldots + (-1)^{n+1}/(n+1)!$. Prove that (a_n) is Cauchy, so tends to some limit l. Prove that each a_n is rational, but l is not.

10

The real numbers as a complete ordered field

In this chapter we shall show how to reverse the process used in the previous chapter. There we postulated the existence of a set satisfying the basic properties (N1)–(N3) of natural numbers, and eventually constructed a set $\mathbb{R}$ which is a complete ordered field. Here we start by postulating the existence of a complete ordered field, and work down until we reach the natural numbers. This approach is basically simpler from a technical point of view (for example, we really do get $\mathbb{N} \subseteq \mathbb{Z} \subseteq \mathbb{Q} \subseteq \mathbb{R}$ without any fudging with isomorphisms).

We begin with some examples of fields, rings, ordered fields, and ordered rings, to show how wide a variety of structures of these types exist. Any system which obeys a formal set of axioms is called a *model* for those axioms, and the power of the axiomatic method is that any deduction from the axioms is true in any model. So any valid deduction from the axioms for an ordered field will hold in the models $\mathbb{Q}$, $\mathbb{R}$ constructed in the last chapter—and indeed in *any* system satisfying the axioms. So we only need to perform the deductions *once*, rather than over again for each model.

The axiomatic method can have another kind of power: the ability to single out (up to isomorphism) a *unique* model. For example, this happened with axioms (N1)–(N3): any systems satisfying them are order isomorphic and therefore to all intents and purposes the same. It will be seen that the same holds for the axioms for a complete ordered field: they define a *unique* system, up to order isomorphism. It is therefore permissible to call such a system *the* real numbers. This is the end of our quest: starting from intuitive ideas about points on a line and decimal expansions we have formulated a set of axioms which defines the required system uniquely. (And we can obtain equally simple descriptions of integers and rational numbers at the same time.)

Examples of rings and fields

Not every system of axioms defines a unique structure, even if isomorphisms are permitted. For example $\mathbb{Z}$ and $\mathbb{Q}$ are both rings, but are not isomorphic since $\mathbb{Q}$ is a field and $\mathbb{Z}$ is not. To motivate the narrowing down of possibilities by imposing extra axioms, we introduce some further examples.

EXAMPLE 1. $\mathbb{Z}_n$, the ring of integers modulo n. Let n be an integer >0, and for $r, s \in \mathbb{Z}$ define

$$r \sim s \Leftrightarrow r - s = kn \text{ for some } k \in \mathbb{Z}.$$

It is easy to show that this is an equivalence relation, and we call the set of equivalence classes $\mathbb{Z}_n$. We denote the class containing m by m_n. The division algorithm extends to give the following statement: if $m, n \in \mathbb{Z}$ and $n > 0$ then there exist $q, r \in \mathbb{Z}$ with $0 \leqslant r < n$, such that $m = qn + r$. Thus $m - r = qn$, so every integer is equivalent to an integer r with $0 \leqslant r < n$. The elements of $\mathbb{Z}_n$ are $0_n, 1_n, \ldots, (n-1)_n$. As in chapter 4 (where we treated the special case $n = 3$) we can define operations on the equivalence classes by

$$r_n + k_n = (r + k)_n,$$
$$r_n k_n = (rk)_n.$$

These operations are well defined and satisfy the axioms for a *ring* whose zero element is 0_n, and unity 1_n.

If n is not prime, then $\mathbb{Z}_n$ is not a field. For if $n = rk$ with $0 < r < n$, $0 < k < n$, then

$$r_n k_n = n_n = 0_n.$$

Say that an element x of a ring is a *zero-divisor* if $x \neq 0$, but $xy = 0$ for some $y \neq 0$, y in the ring. Then r_n and k_n are zero-divisors. But a field has no zero-divisors, for if $xy = 0$ in a field, $y \neq 0$, then we have $x = xyy^{-1} = 0y^{-1} = 0$. Thus $\mathbb{Z}_n$, for composite n, is not a field.

For instance, in $\mathbb{Z}_6$ we have $2_6 3_6 = 0_6$, where $2_6 \neq 0_6$, $3_6 \neq 0_6$. These elements do not have multiplicative inverses in $\mathbb{Z}_6$, as we can check bare-handed just by trying all six possibilities. Thus $2_6 0_6 = 0_6$, $2_6 1_6 = 2_6$, $2_6 2_6 = 4_6$, $2_6 3_6 = 0_6$, $2_6 4_6 = 2_6$, $2_6 5_6 = 4_6$. Nowhere do we get an answer 1_6.

However, if n is *prime*, then $\mathbb{Z}_n$ *is* a field. There are several ways to see this, of which the following is the least sophisticated but the most direct. Given $r_n \neq 0_n$, we look for an inverse by calculating all the products

$$r_n 0_n = 0_n, r_n 1_n = r_n, \ldots, r_n(n-1)_n = ?.$$

All of these elements are different, for if

$$r_n k_n = r_n l_n$$

where $0 \leq k < l < n$, then

$$r_n(l-k)_n = 0_n$$

so that n divides $r(l-k)$. But each factor lies between 0 and n, and n is prime, a contradiction.

Now this list of products contains exactly n elements, all different; and since $\mathbb{Z}_n$ only *has* n elements, each must occur precisely once. In particular 1_n occurs somewhere, say at $r_n k_n = 1_n$; and now k_n is the required inverse. Hence $\mathbb{Z}_n$ is a field if n is prime.

For instance, in $\mathbb{Z}_5$, we look for an inverse for 3_5 by working out $3_5 0_5 = 0_5$, $3_5 1_5 = 3_5$, $3_5 2_5 = 1_5$, $3_5 3_5 = 4_5$, $3_5 4_5 = 2_5$, and the products are precisely the elements of $\mathbb{Z}_5$ in the order $0_5, 3_5, 1_5, 4_5, 2_5$. Among them is 1_5, and the inverse of 3_5 is 2_5.

EXAMPLE 2. $\mathbb{Q}(\sqrt{2}) = \{a + b\sqrt{2} \in \mathbb{R} | a, b \in \mathbb{Q}\}$. This is a field with zero element $0 + 0\sqrt{2}$ and unity $1 + 0\sqrt{2}$. The additive inverse of $a + b\sqrt{2}$ is $-a - b\sqrt{2}$, and if $a + b\sqrt{2} \neq 0$, its multiplicative inverse is

$$\frac{1}{a + b\sqrt{2}} = \frac{a - b\sqrt{2}}{(a + b\sqrt{2})(a - b\sqrt{2})} = \frac{a}{a^2 - 2b^2} + \frac{(-b)}{a^2 - 2b^2}\sqrt{2}.$$

(it is an easy exercise to show that if one of a, b is not 0, then $a^2 - 2b^2 \neq 0$; in fact it is the same as proving $\sqrt{2}$ irrational.)

EXAMPLE 3.   This is going to provide a useful counterexample later. It is: the field $\mathbb{R}(t)$ of *rational functions* in an indeterminate t. An element of $\mathbb{R}(t)$ is most easily described as a quotient of two polynomials

$$\frac{a_n t^n + \ldots + a_0}{b_m t^m + \ldots + b_0}$$

where $a_0, \ldots, a_n, b_0, \ldots, b_m \in \mathbb{R}$ and not all b's are zero. We can think of such an expression as giving rise to a function $f: D \to \mathbb{R}$ where

$$D = \{\alpha \in \mathbb{R} | b_m \alpha^m + \ldots + b_0 \neq 0\}$$

and

$$f(\alpha) = \frac{a_n\alpha^n + \ldots + a_0}{b_m\alpha^m + \ldots + b_0}.$$

This is a quotient of polynomials in the same way that a rational number is a quotient of integers, hence the name 'rational function'.†

The sum and product of rational functions are defined in the customary fashion, and the resulting structure is a field. We may identify $\mathbb{R}$ with the subset of $\mathbb{Q}(t)$ consisting of functions $a_0/1$, where $a_0 \in \mathbb{R}$.

It would be possible to exhibit many other interesting rings and fields: however, those listed above are especially pertinent to this chapter.

Examples of ordered rings and fields

Next we try to introduce orderings. This leads to interesting results when applied to the above examples.

(Non-)example 1. $\mathbb{Z}_n$ cannot be ordered in a way that makes it into an ordered ring. (Of course it can be *ordered* in a way which does not fit the arithmetic, for example,

$$0_n < 1_n < 2_n < \ldots < (n-1)_n.$$

But this does not lead to an ordered ring, for $1_n > 0_n$, $(n-1)_n > 0_n$ would then imply $0_n = 1_n + (n-1)_n > 0_n$, which is absurd.)

More generally, suppose we *could* give $\mathbb{Z}_n$ an order relation making it into an ordered ring. Then there would be a subset $\mathbb{Z}_n^+$ of positive elements, satisfying axioms (O1)–(O3). By (O2) either $1_n \in \mathbb{Z}_n^+$ or $-1_n \in \mathbb{Z}_n^+$. Since $1_n = 1_n.1_n = (-1_n)(-1_n)$, either possibility implies $1_n \in \mathbb{Z}_n^+$ using (O1). Using (O1) and induction we get $2_n = 1_n + 1_n \in \mathbb{Z}_n^+$,

---

† A formal definition of $\mathbb{R}(t)$ can be given as follows. First, a polynomial is determined by its coefficients $a_0, \ldots, a_n$, so we can define a *formal polynomial* to be a sequence $s: \mathbb{N}_0 \to \mathbb{R}$ such that for some $n \in \mathbb{N}_0$ we have $s(m) = 0$ for $m > n$. Write $s(m) = s_m$ and denote s by the sequence $(s_0, s_1, \ldots, s_r, \ldots)$ on the understanding that from some point on we have $s_m = 0$. Addition and multiplication are defined by

$$(s_0, s_1, \ldots, s_r \ldots) + (p_0, p_1, \ldots, p_r \ldots) = (s_0 + p_0, s_1 + p_1, \ldots, s_r + p_r \ldots)$$

$$(s_0, s_1, \ldots, s_r \ldots)(p_0, p_1, \ldots, p_r \ldots) = (s_0 p_0, s_0 p_1 + s_1 p_0, \ldots, q_r \ldots)$$

where $q_r = s_0 p_r + s_1 p_{r-1} + \ldots + s_r p_0$.

The sequence $(0, 1, 0, 0, \ldots)$ can be called t, and then it turns out that $(s_0, s_1, \ldots, s_r \ldots) = s_0 + s_1 t + \ldots + s_r t^r + \ldots$, so we recover the usual notation, as long as we identify $s \in \mathbb{R}$ with the sequence $(s, 0, 0, \ldots)$. The formal polynomials constitute a ring. Using equivalence classes of ordered pairs, in exactly the way we constructed $\mathbb{Q}$ from $\mathbb{Z}$, we then construct $\mathbb{R}(t)$ from the ring of formal polynomials.

$3_n \in \mathbb{Z}_n^+, \ldots, (n-1)_n \in \mathbb{Z}_n^+$. But this leads to the same contradiction as before. Hence $\mathbb{Z}_n$ cannot be given the structure of an ordered ring.

EXAMPLE 2. $\mathbb{Q}(\sqrt{2})$ can be given the structure of an ordered field in two different ways!

The first way is to note that $\mathbb{Q}(\sqrt{2}) \subseteq \mathbb{R}$, and to restrict the usual order relation on $\mathbb{R}$ to $\mathbb{Q}(\sqrt{2})$: clearly this gives $\mathbb{Q}(\sqrt{2})$ the structure of an ordered field.

The second way is more subtle. There is a map $\theta \colon \mathbb{Q}(\sqrt{2}) \to \mathbb{Q}(\sqrt{2})$ defined by

$$\theta(a + b\sqrt{2}) = a - b\sqrt{2}.$$

Now θ is an isomorphism from $\mathbb{Q}(\sqrt{2})$ to itself (usually called an *automorphism*), that is, θ is a bijection, and for all $x, y \in \mathbb{Q}(\sqrt{2})$

$$\theta(x + y) = \theta(x) + \theta(y)$$

$$\theta(xy) = \theta(x)\theta(y).$$

(Check this!) Denoting the first order relation, defined above, by $\geqslant$, we define a new relation $\succcurlyeq$ by

$$x \succcurlyeq y \Leftrightarrow \theta(x) \geqslant \theta(y).$$

The reader should check that this, too, gives $\mathbb{Q}(\sqrt{2})$ the structure of an ordered field. For example, if $x, y \succcurlyeq 0$ then $\theta(x), \theta(y) \geqslant \theta(0) = 0$, so $\theta(x)\theta(y) \geqslant 0$, so $\theta(xy) \geqslant 0$, so $xy \succcurlyeq 0$. The remaining axioms are proved in the same way. Note that in this ordering $\sqrt{2} \prec 0$.

REMARK.   This example should be offset against the following fact: $\mathbb{Z}$ and $\mathbb{Q}$ can be given the structure of ordered rings (or ordered field in the case of $\mathbb{Q}$) in *only one* way. Here is a quick sketch of the reasoning. As we argued for $\mathbb{Z}_n^+$, we always have $1 > 0$ because $1 = 1^2 = (-1)^2$. Inductively it follows that the ordering on $\mathbb{Z}$ must have all natural numbers positive, hence using (O2) the *usual* negative integers must be negative in the given ordering. So for $\mathbb{Z}$, only the usual ordering works. Since everything in $\mathbb{Q}$ is a quotient of integers, the same goes for $\mathbb{Q}$ (after a little work).

The same holds for $\mathbb{R}$, but the proof requires the fact that every positive real number has a square root, a fact which needs verifying. It is an easy consequence of completeness. If $x \in \mathbb{R}$ and $x > 0$, let

$$L = \{y \in \mathbb{R} \mid y > 0 \ \& \ y^2 < x\}.$$

Then L is easily seen to be bounded above and non-empty. By completeness L has a least upper bound u; a quick contradiction argument shows that $u^2 = x$.

Now, for *any* ordering making $\mathbb{R}$ an ordered field, all elements of the form y^2 ($y \in \mathbb{R}$) must be positive, and all elements $-y^2$ must be negative. By what we have just said, the positive and negative elements of $\mathbb{R}$ (in the usual sense) must also be positive and negative respectively in any other ordering, since they are precisely the elements in the required forms. Thus only one ordering exists making $\mathbb{R}$ into an ordered field.

EXAMPLE 3.   We can give the field of rational functions $\mathbb{R}(t)$ an ordering with interesting properties. (This does not give a notion of *size* to a function, but it does not prevent us from imposing an ordering which satisfies axioms (01)–(03).) Define

$$\mathbb{R}(t)^+ = \{f(t) \in \mathbb{R}(t) | \exists \alpha_0 \in \mathbb{R}: \alpha \geqslant \alpha_0 \Rightarrow f(\alpha) \geqslant 0\},$$

which means that $f(t)$ is considered positive if and only if $f(\alpha)$ is positive for all sufficiently large α. (For instance $(t^2 - 17)/(5t^3 + 4t)$ is positive in this sense, but $(t+1)/(3t - t^2)$ is negative.) This, it may be verified, makes $\mathbb{R}(t)$ into an ordered field. If we identify $\mathbb{R}$ with the set of constant functions, as above, then the ordering on $\mathbb{R}(t)$ restricts to the usual ordering on $\mathbb{R}$.

Surprisingly, $\mathbb{R} \subseteq \mathbb{R}(t)$ is *bounded above*. In fact the function $f(t) = t$ is an upper bound. For if $\beta \in \mathbb{R}$ then the function $g(t) = f(t) - \beta = t - \beta$ has the property that $g(\alpha) > 0$ for all $\alpha > \beta$, hence $g(t) \in \mathbb{R}(t)^+$. This proves that t is an upper bound for $\mathbb{R}$ in $\mathbb{R}(t)$.

Isomorphisms again

We have already made use of the concepts 'isomorphism' and 'order isomorphism' in special cases, and it now behoves us to discuss them in general. Recall that if R, S are rings then $\theta: R \to S$ is an *isomorphism* if it is a bijection and if for all r, $s \in R$ we have

$$\theta(r+s) = \theta(r) + \theta(s),$$
$$\theta(rs) = \theta(r)\theta(s). \tag{*}$$

Various axiomatic structures have been proved unique *up to isomorphism*. The reader may wonder why we can't do better, and actually make them *unique*. The reason is that this is just too much to ask, and, in any case, presents no real advantages. An isomorphism, after all, is just a change of name (from r to $\theta(r)$); so given a ring R, we can find lots of isomorphic rings by finding lots of ways changing the names. In formal terms, let S be *any* set for which there exists a bijection $\theta : R \to S$ (we don't assume S is a ring) and use (*) back-to-front, to define ring operations on S, by

$$\theta(r) + \theta(s) = \theta(r+s)$$

$$\theta(r)\theta(s) = \theta(rs).$$

Then S will be isomorphic to R.

How do we know sets S with suitable bijections exist? Take any element t whatever, and let $S = R \times \{t\}$; define θ by $\theta(r) = (r, t)$. This is always a bijection; and different choices of t lead to different choices of S. This shows how wide a variety of sets S can be found; and this is just *one* very simple way to find them.

Since it is the algebraic operations on a ring which are important, and not the elements themselves, an isomorphic ring is just as good as the ring we start from. So it is too restrictive to expect to specify an algebraic structure uniquely; on the other hand uniqueness up to isomorphism is the most that we ever require.

The same goes for an *order isomorphism* between two ordered rings R and S, which in addition to (*) satisfies the condition

$$r \geqslant s \Rightarrow \theta(r) \geqslant \theta(s).$$

We cease philosophizing in order to point out some useful, simple consequences of (*).

LEMMA 1. If $\theta : R \to S$ is an isomorphism of rings, then for all $r \in R$,

(a) $\theta(0) = 0$
(b) $\theta(1) = 1$
(c) $\theta(-r) = -\theta(r)$
(d) $\theta(1/r) = 1/\theta(r)$ provided $1/r$ exists.

PROOF.   For all $r \in R$, we have $r = 0 + r$. Applying θ,

$$\theta(r) = \theta(0+r) = \theta(0) + \theta(r).$$

Now θ is onto, so every element of S is of the form $\theta(r)$ for some $r \in R$. Hence

$$s = \theta(0) + s$$

for all $s \in S$, so by proposition 1 of chapter 9, $\theta(0) = 0$. This proves (a), and (b) is similar. To prove (c),

$$r + (-r) = 0$$

so

$$\theta(r) + \theta(-r) = \theta(0) = 0.$$

By proposition 2 of chapter 9, $\theta(-r) = -\theta(r)$. This proves (c), and (d) is similar. □

If R is a ring, then a *subring* of R is a subset S such that

(i) $r, s \in S \Rightarrow r + s \in S$
(ii) $r, s \in S \Rightarrow rs \in S$
(iii) $s \in S \Rightarrow -s \in S$
(iv) $1 \in S$.

From (iv), (iii), (i) it also follows that $0 = 1 + (-1) \in S$. For example $\mathbb{Z}$ is a subring of $\mathbb{Q}$, and $\mathbb{Q}$ is a subring of $\mathbb{R}$.

As with isomorphisms between rings, it is often sufficient to have a subring isomorphic to something, instead of actually being that thing.

If R is a field, then a *subfield* S is a subring which further satisfies

(v) $s \in S, s \neq 0 \Rightarrow s^{-1} \in S$.

For example, $\mathbb{Q}$ is a subfield of $\mathbb{R}$.
The ideas are applied in the next section.

Some characterizations

PROPOSITION 2.   Every ring R contains a subring isomorphic either to $\mathbb{Z}$ or $\mathbb{Z}_n$ for some n.

PROOF.   Define $\theta: \mathbb{Z} \to R$ by $\theta(0) = 0$, $\theta(1) = 1$, $\theta(n+1) = \theta(n) + 1$ (using the recursion theorem!) for $n > 0$, then let $\theta(-n) = -(\theta(n))$ for $n > 0$. An induction argument shows that

$$\theta(m + n) = \theta(m) + \theta(n)$$

$$\theta(mn) = \theta(m)\theta(n).$$

If θ is an injection, we've finished, for then $\theta(\mathbb{Z})$, the image of $\mathbb{Z}$ under θ, is a subring isomorphic to $\mathbb{Z}$. Unfortunately, θ may not be injective. In this case we argue as follows.

There exist $r > s \in \mathbb{Z}$ such that $\theta(r) = \theta(s)$. Therefore $\theta(r - s) = \theta(r) - \theta(s) = 0$. Using the well ordering property, let n be the smallest natural number such that $n \neq 0$, $\theta(n) = 0$. It follows that $\theta(0)$, $\theta(1), \ldots, \theta(n-1)$ are all different, for if $\theta(r) = \theta(s)$ with $0 < r < s < n$, then $\theta(s - r) = 0$ and this contradicts the definition of n. Also, if $u - v = qn$ $(u, v, q \in \mathbb{Z})$,   then $\theta(u) - \theta(v) = \theta(u - v) = \theta(qn) =$

$\theta(q)\theta(n) = \theta(q)0 = 0$. Hence, using our notation in $\mathbb{Z}_n$, if $u_n = v_n$ then $\theta(u) = \theta(v)$.

We may therefore define a map $\phi: \mathbb{Z}_n \to R$ by $\phi(u_n) = \theta(u)$. The previous remark shows that ϕ is well defined. Now

$$\phi(u_n + v_n) = \phi((u + v)_n) = \theta(u + v) = \theta(u) + \theta(v) = \phi(u_n) + \phi(v_n),$$

$$\phi(u_n v_n) = \phi((uv)_n) = \theta(uv) = \theta(u)\theta(v) = \phi(u_n)\phi(v_n).$$

Since $\theta(0), \ldots, \theta(n - 1)$ are all different, $\phi(0_n), \ldots, \phi((n - 1)_n)$ are all different, so ϕ is an injection. Thus $\phi(\mathbb{Z}_n)$ is a subring of R isomorphic to $\mathbb{Z}_n$. $\square$

For fields we get a similar result:

PROPOSITION 3. Every field F contains a subfield isomorphic either to $\mathbb{Q}$ or to $\mathbb{Z}_p$ (p prime).

PROOF. Using proposition 2, F contains a subring S isomorphic to $\mathbb{Z}$ or to $\mathbb{Z}_n$.

Suppose S is isomorphic to $\mathbb{Z}$, with $\theta: \mathbb{Z} \to S$ an isomorphism. Define $\phi: \mathbb{Q} \to F$ by

$$\phi(m/n) = \theta(m)/\theta(n) \qquad (m, n \in \mathbb{Q}, n \neq 0).$$

Notice that $n \neq 0 \Rightarrow \theta(n) \neq 0$, since θ is injective, so the right-hand side makes sense. Now ϕ is injective, for if $\phi(m/n) = \phi(r/s)$ then

$$\theta(m)/\theta(n) = \theta(r)/\theta(s)$$

so

$$\theta(ms) = \theta(m)\theta(s) = \theta(r)\theta(n) = \theta(rn),$$

hence

$$ms = rn$$

and therefore

$$m/n = r/s.$$

It is now easy to check that $\phi(\mathbb{Q})$ is a subfield isomorphic to $\mathbb{Q}$.

Now suppose S is isomorphic to $\mathbb{Z}_n$. If n is composite, $n = qr$, then $\phi(q_n), \phi(r_n)$ will be zero-divisors in F. But a field F has no zero-divisors ($xy = 0$ if $y \neq 0 \Rightarrow x = xyy^{-1} = 0y^{-1} = 0$). Therefore n is a prime, say $n = p$; and since $\mathbb{Z}_p$ is a field we have found a subfield of F isomorphic to $\mathbb{Z}_p$. $\square$

Next we bring in the order relation.

PROPOSITION 4.   Every ordered ring contains a subring order isomorphic to $\mathbb{Z}$.

PROOF.   By proposition 2 it contains a subring isomorphic to $\mathbb{Z}$ or to $\mathbb{Z}_n$. The proof that $\mathbb{Z}_n$ cannot be made an ordered ring also shows that it cannot be a subring of an ordered ring. The proof that the ordering on $\mathbb{Z}$ is unique shows that the subring isomorphic to $\mathbb{Z}$ is also *order* isomorphic to $\mathbb{Z}$.   □

Similarly we have

PROPOSITION 5.   Every ordered field contains a subfield order isomorphic to $\mathbb{Q}$.

PROOF.   Eliminate the possibility $\mathbb{Z}_p$ as in proposition 4, and then use uniqueness of the order on $\mathbb{Q}$.   □

These two propositions give simple axiomatic characterizations of $\mathbb{Z}$ and $\mathbb{Q}$:
$\mathbb{Z}$ is a minimal ordered ring
(i.e. $\mathbb{Z}$ is an ordered ring with no proper subring);
  $\mathbb{Q}$ is a minimal ordered field
(i.e. $\mathbb{Q}$ is an ordered field with no proper subfield).
These define $\mathbb{Z}$ and $\mathbb{Q}$ uniquely up to isomorphism. For by proposition 4, any minimal ordered ring must be isomorphic to $\mathbb{Z}$, and by proposition 5 any minimal ordered field must be isomorphic to $\mathbb{Q}$.

Finally we turn to complete ordered fields. To deal with these we must extend to them notions such as 'limit' and 'Cauchy sequence'. Thus let F be an ordered field. By proposition 5 it contains a subfield order isomorphic to $\mathbb{Q}$, and by change of notation we may assume without loss of generality that this subfield is $\mathbb{Q}$ itself. We say that a sequence (a_n) of elements of F is *Cauchy* if:
for every $\varepsilon > 0$, $\varepsilon \in F$, there exists $N \in \mathbb{N}_0$ such that for $m, n > N$,

$$|a_m - a_n| < \varepsilon.$$

The sequence (a_n) tends to a *limit* $\lambda \in F$ if for every $\varepsilon > 0$, $\varepsilon \in F$, we can find $N \in \mathbb{N}_0$ such that for all $n > N$

$$|a_n - \lambda| < \varepsilon.$$

As before we write

$$\lim_{n \to \infty} a_n = \lambda, \text{ or } \lim a_n = \lambda.$$

The key result is:

LEMMA 6.   In a complete ordered field, every Cauchy sequence has a limit.

PROOF.   Let (a_n) be a Cauchy sequence in F. By the argument of lemma 12 of chapter 9 (carried out in F) the sequence is bounded. Hence so is every subset of elements in the sequence. Define

$$b_N = \text{the least upper bound of } \{a_N, a_{N+1}, a_{N+2}, \ldots\}.$$

This exists by completeness. Clearly

$$b_0 \geqslant b_1 \geqslant b_2 \geqslant \ldots$$

and the sequence (b_n) is bounded below (say, by any lower bound for (a_n)). Hence we can define

$$c = \text{the greatest lower bound of } (b_n).$$

We claim that c is the limit of the original sequence (a_n).

To prove this, let $\varepsilon > 0$. Suppose that there exist only finitely many values of n with

$$c - \tfrac{1}{2}\varepsilon < a_n < c + \tfrac{1}{2}\varepsilon.$$

Then we may choose N such that for all $n > N$,

$$a_n \leqslant c - \tfrac{1}{2}\varepsilon \quad \text{or} \quad a_n \geqslant c + \tfrac{1}{2}\varepsilon.$$

But there exists $N_1 > N$ such that if $m, n > N_1$ then $|a_m - a_n| < \tfrac{1}{2}\varepsilon$. Hence

$$\text{for all } n > N_1, a_n \leqslant c - \tfrac{1}{2}\varepsilon,$$

or

$$\text{for all } n > N_1, \quad a_n \geqslant c + \tfrac{1}{2}\varepsilon.$$

The latter condition implies that there exists some m with $a_n > b_m$ for all $n > N_1$, which contradicts the definition of b_m. But the former implies that we may change b_{N_1} to $b_{N_1} - \tfrac{1}{2}\varepsilon$, which again contradicts the definition of b_{N_1}.

It follows that for any M there exists $m > M$ such that

$$c - \tfrac{1}{2}\varepsilon < a_m < c + \tfrac{1}{2}\varepsilon.$$

Since (a_n) is Cauchy, there exists $M_1 > M$ such that $|a_n - a_m| < \tfrac{1}{2}\varepsilon$ for $m, n > M_1$. Hence for $n > M_1$,

$$c - \varepsilon < a_n < c + \varepsilon.$$

But this implies that $\lim a_n = c$ as claimed. $\square$

The next step is:

LEMMA 7.   Let $F \supseteq \mathbb{Q}$ be a complete ordered field. If $x \in F$ then there exists $p \in \mathbb{Z}$ such that $p - 1 \leqslant x < p$.

PROOF.   Suppose $n \leqslant x$ for all $n \in \mathbb{Z}$. Then $\mathbb{Z}$ is bounded above by x, so by completeness has a least upper bound k. Hence $n + 1 \leqslant k$ for all $n \in \mathbb{Z}$, because also $n + 1 \in \mathbb{Z}$. This implies $n \leqslant k - 1$, so $k - 1$ is a smaller upper bound for $\mathbb{Z}$. This contradicts the definition of k. Therefore $x < n$ for some $n \in \mathbb{Z}$. Similarly $m < x$ for some $m \in \mathbb{Z}$. Since there are only finitely many integers between m and n we can find an integer p which is the smallest such that $x < p$. Then $p - 1 \leqslant x < p$, and the lemma is proved. $\square$

As a final preparatory step:

LEMMA 8.   Let F be a complete ordered field, and let (a_n) and (b_n) be two sequences with limits a and b respectively. Then

(a) $\lim (a_n + b_n) = a + b$
(b) $\lim (a_n b_n) = ab$.

PROOF.   For the first part, copy the proof of the first theorem in chapter 7 and check that formally it still makes sense. For the second, note that by the argument of lemma 12, chapter 9, we have for all $n \in \mathbb{N}_0$, $|a_n| < A$, $|b_n| < B$ for some $A, B \in F$. Then if $\varepsilon > 0$ we have $\varepsilon / (A + B) > 0$. Hence there exists N_1 such that for $n > N_1$,

$$|a_n - a| < \varepsilon / (A + B)$$

and there exists N_2 such that for $n > N_2$

$$|b_n - b| < \varepsilon / (A + B).$$

Hence for $n > N = \max (N_1, N_2)$,

$$|a_n b_n - ab| = |(a_n - a)b_n + a(b_n - b)|$$
$$< (\varepsilon / (A + B))B + A\varepsilon / (A + B)$$
$$= \varepsilon.$$

This proves (b). (Note the resemblance to the argument just before theorem 15 of chapter 9.) $\square$

For $\mathbb{R}$ we get an even stronger statement than propositions 4 and 5.

THEOREM 9.   Every complete ordered field is order isomorphic to $\mathbb{R}$.

PROOF.   Let F be a complete ordered field. By proposition 5 it has a subfield order isomorphic to $\mathbb{Q}$. For notational convenience as usual we identify this subfield with $\mathbb{Q}$, so that without loss of generality $\mathbb{Q} \subseteq F$.

Elements of $\mathbb{R}$ are equivalence classes $[a_n]$ of Cauchy sequences (a_n) of rationals. We define a map $\theta \colon \mathbb{R} \to F$ by

$$\theta([a_n]) = \lim_{n \to \infty} a_n.$$

First we need to check this makes sense. So we prove (a_n) is Cauchy *in* F (which is not quite the same thing as being Cauchy in $\mathbb{Q}$, because we have more values of ε to look at.) Let $\varepsilon \in F$, $\varepsilon > 0$. We claim that there is a rational ε' with $0 < \varepsilon' < \varepsilon$. Now $1/\varepsilon \in F$ and, by lemma 7, $1/\varepsilon < p$ for some $p \in \mathbb{Z}$. Then $p > 0$, and $0 < 1/p \in \mathbb{Q}$. We take $\varepsilon' = 1/p \in \mathbb{Q}$. Now because (a_n) is Cauchy in $\mathbb{Q}$ it follows that there exists $N \in \mathbb{N}_0$ such that for all $m, n > N$

$$|a_m - a_n| < \varepsilon'.$$

Hence for all $m, n > N$,

$$|a_m - a_n| < \varepsilon,$$

and (a_n) is Cauchy in F. By lemma 6, $\lim a_n$ exists in F.

It is easy to see, using similar arguments, that θ is well defined and injective; and lemma 8 proves that $\theta(\mathbb{R})$ is a subring of F isomorphic to $\mathbb{R}$. It is easy to check that θ preserves the order relation.

It remains to prove that θ is surjective. Let $x \in F$. By lemma 7 there exists an integer a_0 with $a_0 \leqslant x < a_0 + 1$. Inductively (and using lemma 7) we can find integers a_i between 0 and 9 such that

$$a_0 + \frac{a_1}{10} + \ldots + \frac{a_n}{10^n} \leqslant x < a_0 + \frac{a_1}{10} + \ldots + \frac{a_n + 1}{10^n}.$$

Then if $b_n = a_0 + \dfrac{a_1}{10} + \ldots + \dfrac{a_n}{10^n}$ we have

$$|b_n - x| < 1/10^n$$

and it follows easily (using a similar argument to that in the second paragraph of this proof) that

$$\lim b_n = x$$

Also (b_n) is Cauchy in $\mathbb{Q}$, hence $[b_n] \in \mathbb{R}$, and

$$\theta([b_n]) = \lim b_n = x.$$

Therefore θ is surjective, and the theorem is proved.   □

The connection with intuition

We can now tidy our ideas up a little more. We have two types of model of the relevant axiom systems: the formal models $\mathbb{N}_0$, $\mathbb{Z}$, $\mathbb{Q}$, $\mathbb{R}$; and our informal models $\mathscr{N}_0$, $\mathscr{Z}$, $\mathscr{Q}$, $\mathscr{R}$. Now we gave a plausible intuitive explanation of the fact that $\mathscr{R}$ is a complete ordered field, so on this intuitive level theorem 9 tells us that $\mathscr{R}$ and $\mathbb{R}$ are isomorphic. That is to say, the formal construction vindicates our intuition, and can be used to justify all of the properties that we find $\mathscr{R}$ has. Really, it now becomes less serious whether we use the informal $\mathscr{R}$ or the formal $\mathbb{R}$: the work we have done renders both equally safe, and there is now no essential difference between them.

In fact, this chapter and the last show between them that we can build up the number systems either by
(a) postulating the existence of $\mathbb{N}_0$ and constructing $\mathbb{Z}$, $\mathbb{Q}$, $\mathbb{R}$ in turn; or
(b) postulating the existence of $\mathbb{R}$ and constructing $\mathbb{Q}$, $\mathbb{Z}$, $\mathbb{N}_0$ in turn.

By judicious combination of the two methods we can therefore start *anywhere*, such as $\mathbb{Z}$, or $\mathbb{Q}$, and obtain the remaining systems by using chapter 9 to work upwards and this one to work downwards. And the uniqueness theorems proved in this chapter show that it makes no essential difference what method we use: the results are always isomorphic and agree with our intuitive ideas. Precisely where we start has now become a matter of taste rather than a matter of urgency. From any of the different starting points we can give an equally sound development of number systems, and recover all of the standard results of elementary arithmetic, from an axiomatic basis.

Exercises

1. Write out a full proof of proposition 5.

2. Prove that in an ordered field F,

$$a^2 + 1 > 0 \quad \text{for all } a \in F.$$

Deduce that if the equation $x^2 + 1 = 0$ has a solution in a field, then that field cannot be ordered. Find all the solutions of $x^2 + 1 = 0$ in the fields $\mathbb{Z}_2$, $\mathbb{Z}_3$, $\mathbb{Z}_5$.

3. Use the Euclidean algorithm to show that given $m, n \in \mathbb{N}$, there is a technique for calculating $a, b \in \mathbb{Z}$, such that $am + bn = h$, where h is the highest common factor of m,. Deduce that if m, n are coprime, then there exist integers a, b such that $am + bn = 1$. Find a, b when $m = 1008$, $n = 1375$. Calculate the multiplicative inverse of 1008_{1375} in $\mathbb{Z}_{1375}$.

Show that m_n has a multiplicative inverse in $\mathbb{Z}_n$ if and only if m and n are coprime.

4. In an ordered ring, prove that for all x, y

$$\|x| - |y\| \leqslant |x + y| \leqslant |x| + |y|.$$

5. From the axioms of a complete ordered field, prove that every positive element a of $\mathbb{R}$ has a unique positive square root. (Hint: consider $\{x \in \mathbb{Q} | x^2 \leqslant a\}$.)

6. Prove by induction that $0 \leqslant a \leqslant b \Rightarrow a^n \leqslant b^n$ in an ordered ring R. Given $a \in R$, $a \geqslant 0$, show that if there exists an element $r \in R$ such that $r \geqslant 0$ and $r^n = a$, then it is unique.

7. Show that every positive element in a complete ordered field has a unique nth root. (Consider $\{x | x^n \leqslant a\}$.)

8. Use exercise 7 to define $x^{p/q}$ for a positive element x in a complete ordered field and a rational number p/q.

9. Define a field $\mathbb{Q}(\sqrt{3})$ analogous to $\mathbb{Q}(\sqrt{2})$ and show that there are two different ways of making it into an ordered field.

10. Show that the two orderings mentioned for $\mathbb{Q}(\sqrt{2})$ are the only order relations under which it is an ordered field.

11. Find a field with exactly four different orderings which make it an ordered field.

12. Let $\mathbb{R}[t]$ be the ring of polynomials $p(t) = a_n t^n + a_{n-1} t^{n-1} + \ldots + a_0$ with real coefficients. Define the relation $\geqslant$ by

$$p(t) \geqslant q(t) \Leftrightarrow p(0) \geqslant q(0).$$

Does this make $\mathbb{R}[t]$ into an ordered ring?

13. In any ordered field F, we can define the concept of a limit by

$$a_n \to a \text{ as } n \to \infty \text{ means}$$

given $e \in F$, $e > 0$, there exists $N \in \mathbb{N}$ such that $n > N \Rightarrow |a_n - a| < e$. In the field $\mathbb{R}(t)$ of example 3, page 193, show that the sequence $(1/n)$ does *not* tend to zero. (Find $e \in \mathbb{R}(t)$ such that $1/n > e$ for all $n \in \mathbb{N}$.) Show that if F is a *complete* ordered field, then $1/n \to 0$.

14. Let F be an ordered field. Show that F is complete if and only if *both* of the following properties hold:
  (i) every Cauchy sequence converges, (Cauchy completeness),
  (ii) if $e \in F$, $e > 0$, then $1/10^n < e$ for some $n \in \mathbb{N}$, (Archimedes' condition).

15. *Challenge.* Show that Cauchy completeness in an ordered field does not imply completeness.

11

Complex numbers and beyond

COMPLEX numbers are still regarded by some with a mixture of suspicion and awe, but to a modern mathematician they are just a simple set-theoretic extension of the real numbers. In this chapter we will show how to construct them from $\mathbb{R}$, completing the standard hierarchy of number systems $\mathbb{N}_0 \subseteq \mathbb{Z} \subseteq \mathbb{Q} \subseteq \mathbb{R} \subseteq \mathbb{C}$. We could go on to look for an extension of $\mathbb{C}$. The nineteenth century mathematician Hamilton found one which he named the quaternions which we'll briefly describe. But the moral of modern mathematics is that we must broaden our horizons and look to axiomatic systems which describe more useful mathematical structures. The concept of number is but a part of this general study. Modern algebra concerns itself with axiomatic systems which, broadly speaking, consist of sets with various operations on them. We've already met two, namely rings and fields, but there are many others. This is not an algebra book, so we won't study any of them in detail, but it's worth mentioning the important ones. Looking beyond complex numbers, the more fruitful direction is not towards Hamilton's quaternions, but to the generalized algebraic structures of modern algebra.

Historical background

We've already mentioned in chapter 1 the problems associated with the acceptance of complex numbers as a genuine concept. It's worth pausing briefly to look at a historical outline, because it may help the reader to unlock misconceptions which often occur.

At the beginning of the sixteenth century there was much interest in solving algebraic equations, one of which was:

Find two numbers whose sum is ten and product is forty.

In modern notation this gives equations

$$x + y = 10$$
$$xy = 40$$

Substituting for y from the first equation into the second, we find

$$x(10-x) = 40,$$

so

$$x^2 - 10x + 40 = 0$$

with solutions

$$x = \frac{+10 \pm \sqrt{(100-160)}}{2} = 5 \pm \sqrt{(-15)}.$$

Corresponding to $x = 5+\sqrt{(-15)}$ if $y = 5-\sqrt{(-15)}$ and vice versa, so the solution of the problem is the pair of expressions $5+\sqrt{(-15)}$, $5-\sqrt{(-15)}$. Sixteenth century mathematicians realized that these were not real numbers. The square of any real number is positive, so -15 is not the square of a real number and $\sqrt{(-15)}$ cannot be real. Nevertheless, manipulating these expressions, no matter what $\sqrt{(-15)}$ is, when we add the solutions, the terms $\pm\sqrt{(-15)}$ cancel, giving

$$(5+\sqrt{(-15)}) + (5-\sqrt{(-15)}) = 10$$

and when we multiply them, then

$$(5+\sqrt{(-15)})(5-\sqrt{(-15)}) = 5^2 - (\sqrt{(-15)})^2$$
$$= 25 - (-15)$$
$$= 40.$$

Simply by treating $\sqrt{(-15)}$ as an 'imaginary' number and manipulating it algebraically as if its square were -15, the expressions $5+\sqrt{(-15)}$, $5-\sqrt{(-15)}$ solve the problem. Now any positive real number a has a positive square root $\sqrt{a}$. The square root of a negative real number $-a$ $(a>0)$, if there were such a thing, could be written $\sqrt{(a)} = \sqrt{(-1)}\sqrt{a}$. The eighteenth century mathematician Euler introduced the symbol† i for $\sqrt{(-1)}$, so that $\sqrt{(-a)} = i\sqrt{a}$. An expression of the form $x+iy$ where $x, y \in \mathbb{R}$ was called a *complex number*, though it was still not clear what this really was. A quadratic equation

$$ax^2 + bx + c = 0 \qquad (a, b, c \in \mathbb{R})$$

now has a solution

$$x = \frac{-b \pm \sqrt{(b^2 - 4ac)}}{2a} \qquad \text{for } b^2 \geqslant 4ac$$

† Modern engineers and some mathematicians now use the symbol j for $\sqrt{(-1)}$.

and

$$x = \frac{-b \pm i\sqrt{(4ac - b^2)}}{2a} \quad \text{for } b < 4ac.$$

In other words if $b^2 \geqslant 4ac$, then the equation has real solutions, but if $b^2 < 4ac$ it does not have real solutions at all, but it does have complex ones.

This set up a dichotomy in the eyes of earlier mathematicians between real (in the sense of genuine) and imaginary (in the sense of non-existent) solutions to equations. Complex numbers were saddled with the psychological overtones associated to the words 'complex' and 'imaginary'.

In 1806 the French mathematician Argand described a complex number $x + iy$ as a point in the plane:

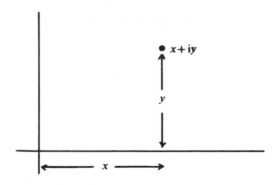

The horizontal axis became the *real axis*, the vertical axis the *imaginary axis*, and the number i was seen as the point one unit up on ˙ maginary axis.

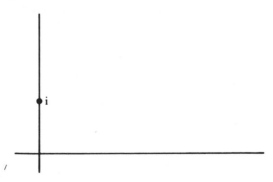

This description was given the name 'Argand diagram' by which name the picture of complex numbers as points in the plane is often known today, although the idea was put forward earlier in the doctoral thesis of the great German mathematician Gauss (1799) and even this was predated by the little known work of a Danish surveyor Wessell (1797)—such are the vagaries of historical acknowledgement. Although the complex numbers were now described concretely as points in the plane, the mystification of earlier eras still shrouded them. Gauss clearly regarded a complex number as a pair (x, y) of real numbers. In the 1830's the Irish mathematician Hamilton canonized them as 'couples of real numbers' (a couple being his name for an ordered pair). This is the heart of the matter and the key to the modern description: a point in the plane is an ordered pair (x, y) and the symbol $x + iy$ is just another name. The mysterious expression i is none other than the ordered pair $(0, 1)$.

Construction of the complex numbers

Let $\mathbb{C}$ be another name† for $\mathbb{R}^2$, the set of ordered pairs (x, y) for $x, y \in \mathbb{R}^2$. Define addition and multiplication on $\mathbb{C}$ by

$$(x_1, y_1) + (x_2, y_2) = (x_1 + x_2, y_1 + y_2) \tag{1}$$

$$(x_1, y_1)(x_2, y_2) = (x_1 x_2 - y_1 y_2, x_1 y_2 + x_2 y_1). \tag{2}$$

It is a simple matter to check that $\mathbb{C}$ is a field under these operations with zero $(0, 0)$, and unity $(1, 0)$. The negative of (x, y) is $(-x, -y)$ and if $(x, y) \neq (0, 0)$, the multiplicative inverse of (x, y) is

$$\left(\frac{x}{x^2 + y^2}, \frac{-y}{x^2 + y^2} \right).$$

Define $f: \mathbb{R} \rightarrow \mathbb{C}$ by $f(x) = (x, 0)$, then

$$f(x_1 + x_2) = (x_1 + x_2, 0) = (x_1, 0) + (x_2, 0) = f(x_1) + f(x_2)$$

and

$$f(x_1 x_2) = (x_1 x_2, 0) = (x_1, 0)(x_2, 0) = f(x_1)f(x_2).$$

The function f is clearly an injection and so is an isomorphism of fields, from $\mathbb{R}$ onto the subfield $f(\mathbb{R}) \subseteq \mathbb{C}$. This subfield $f(\mathbb{R})$ is none other than the 'real axis' of Argand's description.

---

† We often describe $\mathbb{C}$ as the 'complex plane', because of Argand's description, but we now see that as a set, $\mathbb{C}$ is precisely the same as $\mathbb{R}^2$, so, as a set of elements, the complex plane and the real plane are one and the same thing!

As usual, we will consider $\mathbb{R}$ as a subset of $\mathbb{C}$ via this isomorphism, which amounts to regarding the real numbers as the real axis in the complex plane and replacing the symbol $(x, 0)$ by x.

Define i to be the ordered pair $(0, 1)$, then, using (2),

$$i^2 = (0, 1)^2 = (-1, 0).$$

Thinking of $(-1, 0)$ as the real number -1, this gives $i^2 = -1$.

More generally, using (1) and (2)

$$(x, 0) + (0, 1)(y, 0) = (x, 0) + (0, y) = (x, y).$$

Replacing $(x, 0)$, $(y, 0)$ by x, $y \in \mathbb{R}$ respectively, this says

$$x + iy = (x, y).$$

The 'complex number'† $x + iy$ is another name for the ordered pair (x, y).

As a matter of interest, returning to the definitions of addition and multiplication (1), (2) in terms of this notation, we find

$$(x_1 + iy_1) + (x_2 + iy_2) = (x_1 + x_2) + i(y_1 + y_2)$$

$$(x_1 + iy_1)(x_2 + iy_2) = (x_1 x_2 - y_1 y_2) + i(x, x_2 + x_2 y_1).$$

We thus recover the usual addition and multiplication of complex numbers (which is why definitions (1), (2) were set up in the first place!).

Historically in the expression $x + iy$, x is referred to as the 'real part' and y as the 'imaginary part'. Both of x, y are real numbers, being the first and second coordinates of the ordered pair $(x, y) \in \mathbb{R}^2$. If we have

$$x_1 + iy = x_2 + iy_2$$

then

$$(x_1, y_1) = (x_2, y_2),$$

and by the usual properties of ordered pairs,

$$x_1 = x_2, \qquad y_1 = y_2.$$

Historically this deduction was referred to as 'comparing real and

---

† Some people have the misconception that a complex number is $x + iy$ where x, y are real and $y \neq 0$, reserving the name 'real number' for $x + iy$ where $y = 0$. Mathematicians regard *all* expressions $x + iy$ $(x, y \in \mathbb{R})$ as complex numbers, and this *includes* real numbers.

imaginary parts'; we now see it as an application of the set-theoretic definition of ordered pairs.

A modern interpretation of the solution of the quadratic $x^2 - 10x + 40 = 0$ is that there are no solutions in $\mathbb{R}$, but considering this as an equation in $\mathbb{C}$, then there are solutions $5 \pm i\sqrt{15}$. This illustration is no more 'complex' than what happens in considering the equation $2x = 1$ in $\mathbb{N}$ or $\mathbb{Q}$. There is no solution to the equation $2x = 1$ in $\mathbb{N}$, but in $\mathbb{Q}$ there is the solution $x = \frac{1}{2}$.

Time and again in mathematics, we will find a problem doesn't have a solution in a given context, but, interpreted in a wider context, a solution is possible. Don't be surprised by this phenomenon or give it unwarranted mystical significance.

Conjugation

A complex number $x + iy$ may be denoted by a single symbol z (or any other suitable letter for that matter). When we write $z = x + iy$, then we will always suppose $x, y \in \mathbb{R}$, unless something is stated to the contrary. In these circumstances we define the *conjugate* of $z = x + iy$ to be

$$\bar{z} = x - iy.$$

For instance $\overline{3 + 2i} = 3 - 2i$, $\overline{1 - 2i} = 1 + 2i$ and so on. Conjugation has certain elementary properties which we collect together as:

PROPOSITION 1.

(a) $\overline{z_1 + z_2} = \bar{z}_1 + \bar{z}_2$
(b) $\overline{z_1 z_2} = \bar{z}_1 \bar{z}_2$
(c) $\bar{\bar{z}} = z$
(d) $z = \bar{z} \Leftrightarrow z \in \mathbb{R}$.

PROOF.   Elementary checking of definitions.   □

If we define $c \colon \mathbb{C} \to \mathbb{C}$ by $c(z) = \bar{z}$, then proposition 1 tells us that c is an automorphism of the field $\mathbb{C}$ which is the identity when restricted to $\mathbb{R}$.

The modulus

If $z = x + iy$ where $x, y \in \mathbb{R}$, then $x^2 + y^2 \geq 0$. Any positive real number has a unique positive square root. The *modulus* of $z \in \mathbb{C}$ is defined to be

$$|z| = \sqrt{(x^2 + y^2)}.$$

For instance $|3+2i| = \sqrt{(3^2+2^2)} = \sqrt{13}, |-5| = \sqrt{25} = 5$. In particular, for any real number x, $|x| = \sqrt{x^2}$, and, since the positive square root is taken, this reduces to the usual definition of modulus in the real case,

$$|x| = \begin{cases} x & \text{for } x \geqslant 0 \\ -x & \text{for } x < 0 \end{cases} \quad \text{for } x \in \mathbb{R}.$$

In geometric terms, the modulus is the distance from the origin to the point $x + iy$ in the complex plane.

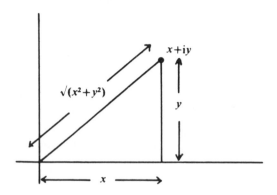

If $z_1 = x_1 + iy_1$, $z_2 = x_2 + iy_2$, then

$$|z_2 - z_1| = |(x_2 - x_1) + i(y_2 - y_1)|$$
$$= \sqrt{((x_2 - x_1)^2 + (y_2 - y_1)^2)}.$$

This is the distance from the point z_1 to the point z_2 in the plane:

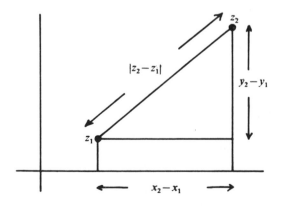

PROPOSITION 2.
(a) $|z| \in \mathbb{R}$, $|z| \geqslant 0$ for all $z \in \mathbb{C}$.
(b) $|z| = 0 \Leftrightarrow z = 0$.
(c) $|z|^2 = z\bar{z}$.
(d) $|z_1 z_2| = |z_1| \, |z_2|$.
(e) $|z_1 + z_2| \leqslant |z_1| + |z_2|$.

PROOF.   (a) and (b) are straightforward.
(c) follows from the definitions, for if $z = x + iy$, then

$$z\bar{z} = (x + iy)(x - iy) = x^2 - (iy)^2 = x^2 + y^2 = |z|^2.$$

(d) Since $|z| \geqslant 0$ for all $z \in \mathbb{C}$, it is sufficient to show

$$|z_1 z_2|^2 = |z_1|^2 |z_2|^2.$$

But

$$
\begin{aligned}
|z_1 z_2|^2 &= (z_1 z_2)(\overline{z_1 z_2}) &&\text{by 2 (c)} \\
&= z_1 z_2 \bar{z}_1 \bar{z}_2 &&\text{by 1 (b)} \\
&= z_1 \bar{z}_1 z_2 \bar{z}_2 \\
&= |z_1|^2 |z_2|^2.
\end{aligned}
$$

(e) A frontal attack on this equality leads to some intricate algebra which can be forced through. We can be more refined, but less direct by writing $z_1 = x_1 + iy_1$, $z_2 = x_2 + iy_2$ and considering the identity

$$(x_1^2 + y_1^2)(x_2^2 + y_2^2) - (x_1 x_2 + y_1 y_2)^2 = (x_1 y_2 - x_2 y_1)^2$$

which immediately tells us that

$$(x_1 x_2 + y_1 y_2)^2 \leqslant (x_1^2 + y_1^2)(x_2^2 + y_2^2) = |z_1|^2 |z_2|^2.$$

Taking square roots, we get

$$x_1 x_2 + y_1 y_2 \leqslant |z_1| \, |z_2|$$

which is valid even if $x_1 x_2 + y_1 y_2$ is negative. Hence

$$2(x_1 x_2 + y_1 y_2) \leqslant 2|z_1| \, |z_2|$$

which gives

$$x_1^2 + 2x_1 x_2 + x_2^2 + y_1^2 + 2y_1 y_2 + y_2^2 \leqslant x_1^2 + y_1^2 + 2|z_1| \, |z_2| + x_2^2 + y_2^2;$$

this simplifies to

$$(x_1+x_2)^2+(y_1+y_2)^2 \leqslant |z_1|^2+2|z_1|\,|z_2|+|z_2|^2,$$

which is

$$|z_1+z_2|^2 \leqslant (|z_1|+|z_2|)^2.$$

Since the modulus is positive, we can take square roots to give

$$|z_1+z_2| \leqslant |z_1|+|z_2|. \quad \square$$

Part (c) of this proposition gives a nice description of the inverse of $z = x+iy$ when $z \neq 0$, for then $|z|^2 = x^2+y^2 \neq 0$, so the equation $z\bar{z} = |z|^2$ implies

$$z\bar{z}/|z|^2 = 1,$$

so

$$z^{-1} = \bar{z}/|z|^2.$$

It is also worth emphasizing that, although part (e) of the proposition involves an inequality, this is between two *real* numbers $|z_1+z_2|$ and $|z_1|+|z_2|$. Although the subfield $\mathbb{R}$ is an ordered field, $\mathbb{C}$ itself is not. That is not to say we cannot order $\mathbb{C}$ in the sense of chapter 4. We certainly can. For instance, we can define the relation $\geqslant$ by

$$x_1+iy_1 \geqslant x_2+iy_2 \Leftrightarrow \text{either } x_1 \geqslant x_2 \quad \text{or} \quad x_1 = x_2 \text{ and } y_1 \geqslant y_2.$$

This is certainly an order relation. However, it does not blend happily with the arithmetic: for instance we have

$$z_1 \geqslant 0, z_2 \geqslant 0 \nRightarrow z_1 z_2 \geqslant 0,$$

as is demonstrated by the example

$$i \geqslant 0, \quad \text{but} \quad i^2 = -1 \ngeqslant 0.$$

There is just no way that we can put an order on $\mathbb{C}$ which fits with the arithmetic on $\mathbb{C}$ such that $\mathbb{C}$ is an ordered field in the sense of chapter 9. This would require us to find a subset $\mathbb{C}^+ \subseteq \mathbb{C}$ such that

(i) $z_1, z_2 \in \mathbb{C}^+ \Rightarrow z_1+z_2 \in \mathbb{C}^+$ and $z_1 z_2 \in \mathbb{C}^+$,
(ii) $z \in \mathbb{C} \Rightarrow z \in \mathbb{C}^+$ or $-z \in \mathbb{C}^+$,
(iii) $z \in \mathbb{C}^+$ and $-z \in \mathbb{C}^+ \Rightarrow z = 0$.

But (ii) gives $i \in \mathbb{C}^+$ or $-i \in \mathbb{C}^+$; in the first case (i) implies $i^2 \in \mathbb{C}^+$, in the second $(-i)^2 \in \mathbb{C}^+$, so in either case $-1 \in \mathbb{C}^+$. Applying (i) again we find $(-1)^2 \in \mathbb{C}^+$, so $1 \in \mathbb{C}^+$. This contradicts (iii) because $1, -1 \in \mathbb{C}^+$ but

$1 \neq 0$. Because of this lack of an order on $\mathbb{C}$, inequalities between complex numbers like $z_1 > z_2$ are nonsense unless the numbers involved are real. A formula like $|z_1| > |z_2|$ is perfectly feasible, because $|z_1|, |z_2| \in \mathbb{R}$ and the real numbers are an ordered field.

Hamilton's quaternions

We might attempt to extend the number system $\mathbb{N}_0 \subseteq \mathbb{Z} \subseteq \mathbb{Q} \subseteq \mathbb{R} \subseteq \mathbb{C}$ further and look for an extension of $\mathbb{C}$. For years in the last century the Irish mathematician Hamilton followed up his conception of complex numbers as ordered couples (x, y) of real numbers, searching for a system of triples (x_1, x_2, x_3) or quadruples (x_1, x_2, x_3, x_4) which extended beyond the complex numbers. In 1843 he found a system of quadruples which is 'almost' a field in the precise sense that it satisfies all the field axioms except commutativity of multiplication.

A *division ring* is an algebraic system consisting of a set D and two binary operations $+, .$ on D such that for all $a, b, c \in D$,

(A1) $(a+b)+c = a+(b+c)$.

(A2) There exists $0 \in D$ such that for all $a \in D$, $0+a = a+0 = a$.

(A3) Give $a \in D$, there exists $-a \in D$ such that $a+(-a) = (-a)+a = 0$.

(A4) $a+b = b+a$.

(M1)† $(ab)c = a(bc)$.

(M2) There exists $1 \in D$, $1 \neq 0$ such that for all $a \in D$, $a1 = 1a = a$.

(M3) Given $a \in D$, $a \neq 0$, there exists $a^{-1} \in D$ such that $aa^{-1} = a^{-1}a = 1$.

(D) $a(b+c) = ab+ac$, $(b+c)a = ba+ca$.

Hamilton's quaternions, as he called his system of quadruples, is an example of a division ring. Its multiplication is not commutative in the sense that for some elements a, b we have $ab \neq ba$.

His discovery can be explained in terms of three symbols i, j, k multiplied according to the rules:

$$i^2 = j^2 = k^2 = -1$$

$$ij = k, \ jk = i, \ ki = j$$

$$ji = -k, \ kj = -i, \ ik = -j.$$

The last six of these can be described by writing the symbols i, j, k in a

---

† As usual we will write ab for $a \cdot b$.

clockwise cycle

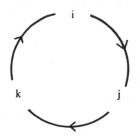

and then the product of any two in clockwise order is the third and the product anticlockwise is minus the third.

Hamilton thought of a quadruple of real numbers (x_1, x_2, x_3, x_4) as $x_1 + ix_2 + jx_3 + kx_4$. He added them in the obvious way:

$$(x_1 + ix_2 + jx_3 + kx_4) + (y_1 + iy_2 + jx_3 + ky_4)$$
$$= (x_1 + y_1) + i(x_2 + y_2) + j(x_3 + y_3) + k(x_4 + y_4),$$

and multiplied them using the above rules for multiplying i, j, k.

Written out in full this amounts to:

$$(x_1 + ix_2 + jx_3 + kx_4)(y_1 + iy_2 + jy_3 + ky_4)$$
$$= x_1 y_1 - x_2 y_2 - x_3 y_3 - x_4 y_4$$
$$+ i(x_1 y_2 + x_2 y_1 + x_3 y_4 - x_4 y_3)$$
$$+ j(x_1 y_3 - x_2 y_4 + x_3 y_1 + x_4 y_2)$$
$$+ k(x_1 y_4 + x_2 y_3 - x_3 y_2 + x_4 y_1)$$

This can be written in terms of ordered quadruples simply by replacing each $a_1 + ia_2 + ja_3 + ka_4$ by (a_1, a_2, a_3, a_4) in the obvious manner.

Formally addition and multiplication of such quadruples is given by

$$(x_1, x_2, x_3, x_4) + (y_1, y_2, y_3, y_4) = (x_1 + y_1, x_2 + y_2, x_3 + y_3, x_4 + y_4)$$

$$(x_1, x_2, x_3, x_4)(y_1, y_2, y_3, y_4) = (a_1, a_2, a_3, a_4)$$

where

$$a_1 = x_1 y_1 - x_2 y_2 - x_3 y_3 - x_4 y_4,$$
$$a_2 = x_1 y_2 + x_2 y_1 + x_3 y_4 - x_4 y_3,$$
$$a_3 = x_1 y_3 - x_2 y_4 + x_3 y_1 + x_4 y_2,$$
$$a_4 = x_1 y_4 + x_2 y_3 - x_3 y_2 + x_4 y_1.$$

We denote the set of all quadruples by $\mathbb{H}$ (for Hamilton). These quadruples are also called *quaternions* or *hypercomplex numbers*.

PROPOSITION 3.   The quaternions $\mathbb{H}$ form a division ring.

PROOF.   This is simply a matter of checking the axioms (A1)–(A4), (M1)–(M3), (D) for $\mathbb{H}$. They are all straightforward, although we will be the first to admit that the associativity of multiplication (M1) is tedious to say the least. The zero element in (A2) is $(0, 0, 0, 0)$, the negative of (x_1, x_2, x_3, x_4) in (A3) is $(-x_1, -x_2, -x_3, -x_4)$, the unity element in (M2) is $(1, 0, 0, 0)$ and the inverse of $(x_1, x_2, x_3, x_4) \neq (0, 0, 0, 0)$ in (M3) is

$$(x_1, x_2, x_3, x_4)^{-1} = (x_1/a, -x_2/a, -x_3/a, -x_4/a)$$

where $a = x_1^2 + x_2^2 + x_3^2 + x_4^2$. $\square$

Multiplication in $\mathbb{H}$ is not generally commutative, for instance

$$(0, 1, 0, 0)(0, 0, 1, 0) = (0, 0, 0, 1)$$

but

$$(0, 0, 1, 0)(0, 1, 0, 0) = (0, 0, 0, -1).$$

Writing $i = (0, 1, 0, 0)$, $j = (0, 0, 1, 0)$, $k = (0, 0, 0, 1)$, then this amounts to $ij = k$, $ji = -k$ as explained previously. Hamilton's other rules for multiplication of i, j, k also follow, simply because we set up the rule of multiplication to make it happen that way

If we look at the subset $C = \{(x, y, 0, 0) \in \mathbb{H} | x, y \in \mathbb{R}\}$, then we find that multiplication on C reduces to

$$(x_1, y_1, 0, 0)(x_2, y_2, 0, 0) = (x_1 x_2 - y_1 y_2, x_1 y_2 + x_2 y_1, 0, 0),$$

and that this is commutative. The map $f: \mathbb{C} \to \mathbb{H}$ given by $f(x + iy) = (x, y, 0, 0)$ is easily seen to be an isomorphism of fields from $\mathbb{C}$ to C. Via this isomorphism we can regard $\mathbb{C}$ as a subset of $\mathbb{H}$. Writing (x_1, x_2, x_3, x_4) as $x_1 + ix_2 + jx_3 + kx_4$ the function $f: \mathbb{C} \to \mathbb{H}$ simply becomes

$$f(x + iy) = x + iy + j0 + k0.$$

The inclusion $\mathbb{C} \subseteq \mathbb{H}$ is then given by regarding the complex number $x + iy$ as the quaternion $x + iy + j0 + k0$.

In many ways properties of $\mathbb{C}$ can be generalized to $\mathbb{H}$ (hence the name 'hypercomplex numbers'). For instance we can define the *conjugate* of a quaternion $q = x_1 + ix_2 + jx_3 + kx_4$ to be

$$\bar{q} = x_1 - ix_2 - jx_3 - kx_4.$$

This has some of the properties of the complex conjugate, but not all. In particular

$$\overline{q_1 + q_2} = \bar{q}_1 + \bar{q}_2$$

$$\bar{\bar{q}} = q$$

$$q = \bar{q} \Leftrightarrow q \in \mathbb{R}.$$

The rule for the conjugate of a product becomes

$$\overline{q_1 q_2} = \bar{q}_2 \bar{q}_1,$$

as the reader may verify by explicit calculation. Because multiplication is not commutative, we can't straighten this out by reversing the order of $\bar{q}_2$ and $\bar{q}_1$.

We can also define the *modulus* of a quaternion $q = x_1 + ix_2 + jx_3 + kx_4$ to be

$$|q| = \sqrt{(x_1^2 + x_2^2 + x_3^2 + x_4^2)}.$$

In this case we find

$$|q| \in \mathbb{R}, \qquad |q| \geqslant 0 \quad \text{for all } q \in \mathbb{H},$$

$$|q| = 0 \Leftrightarrow q = 0.$$

$$q\bar{q} = |q|^2.$$

$$|q_1 q_2| = |q_1|\,|q_2|.$$

$$|q_1 + q_2| \leqslant |q_1| + |q_2|.$$

The proofs of these vary in difficulty and are omitted. (They are analogous to the complex case, taking care over the non-commutativity of $\mathbb{H}$.)

As with complex numbers, for $q \in \mathbb{H}$, $q \neq 0$, we find $q\bar{q} = |q|^2$ where $|q|^2 \neq 0$, so

$$q\bar{q}/|q|^2 = 1$$

and

$$q^{-1} = \bar{q}/|q|^2.$$

Some properties of the quaternions are startling, to say the least. For instance, we know that $i^2 = j^2 = k^2 = (-i)^2 = (-j)^2 = (-k)^2 = -1$, so the equation

$$x^2 + 1 = 0$$

has at least six solutions in $\mathbb{H}$, namely $\pm i$, $\pm j$, $\pm k$.

In fact $(ib + jc + kd)^2 = -b^2 - c^2 - d^2$, so any quaternion $ib + jc + kd$ where $b^2 + c^2 + d^2 = 1$ is a solution of $x^2 + 1 = 0$. There are an *infinite* number of solutions in $\mathbb{H}$.

In $\mathbb{Q}$ the equation $x^2 + 1 = 0$ has no solutions, in $\mathbb{C}$ it has two. Every quadratic equation has at most two solutions in $\mathbb{C}$, but when we pass to $\mathbb{H}$ and lose commutativity of multiplication, we lose this property also.

The quaternions are an interesting algebraic system worthy of study in their own right, but their significance lies historically in the non-commutativity of multiplication. This freed mathematics of the tyranny of the commutative property, whereby the multiplication of numbers had previously been considered commutative as an immutable, preordained law. It was now consistent to have algebraic systems in which 'a times b' need not equal 'b times a', and this led to the many possible generalized algebraic structures studied in modern mathematics. An algebraic structure is now regarded as a set together with binary operations on that set satisfying various properties. Let us briefly look at some of them, so that the reader can see part of the broad spectrum of systems available in the algebra of today.

Semigroups and groups

We begin with a set X and a binary operation $\circ$ on X.

(1) If $(a \circ b) \circ c = a \circ (b \circ c)$ for all $a, b, c \in X$, then $\circ$ is said to be *associative*.
(2) An element $e \in X$ satisfying $a \circ e = e \circ a = a$ for all $a \in X$ is said to be an *identity*.
(3) If an identity e exists, then for $a \in X$, an element $b \in X$ is called an *inverse* for a if $a \circ b = b \circ a = e$.
(4) If $a \circ b = b \circ a$ for all $a, b \in X$, then $\circ$ is said to be *commutative*.

A set X with a binary operation satisfying (1) and (2) is called a *semigroup*; if (3) is also satisfied, it is called a *group*, and if (1)–(4) all hold then it is called a *commutative*, or *abelian, group*.

EXAMPLES
(1) $\mathbb{N}_0$ under the binary operation $+$ is a semigroup with identity 0.
(2) $\mathbb{N}_0$ is a semigroup under multiplication with identity 1.
(3) $\mathbb{Z}$ is a semigroup under multiplication.
(4) $\mathbb{Z}$ is a group under addition. The identity is zero and the inverse of $n \in \mathbb{Z}$ is $-n$ because $n + 0 = 0 + n = n$ and $n + (-n) = (-n) + n = 0$.

(5) The non-zero elements of $\mathbb{H}$ is a group under multiplication. The identity if 1 and the inverse of $q \in \mathbb{H}\backslash\{0\}$ is $\bar{q}/|q|$.

Examples (1)–(4) are commutative, example (5) is non-commutative.

Groups and semigroups occur in many diverse situations in mathematics. For instance if A is a set, the set X of all functions from A to A is a semigroup under composition. Thus $f \in X$ means $f: A \to A$, and given $f, g \in X$, then the composition of f, g, which we write as gf, is just the usual composition of maps, $gf: A \to A$, given by

$$gf(a) = g(f(a)) \quad \text{for all } a \in A.$$

The identity element of X is the identity map $i_A: A \to A$, $i_A(a) = a$. An element $f \in X$ has an inverse if and only if it is a bijection, and then the inverse is just the set-theoretic inverse. Let $B(A)$ be the set of bijections from A to A, so $f \in B(A)$ means '$f: A \to A$ is a bijection'. Then $B(A)$ is a group. If A has at least three elements, then $B(A)$ is not commutative. Let $a, b, c \in A$ be distinct elements and define θ, $\phi \in B(A)$ by

$$\theta(a) = b, \quad \theta(b) = a, \quad \theta(c) = c, \quad \theta(x) = x \quad \text{for all other } x \in A,$$

$$\phi(a) = c, \quad \phi(b) = b, \quad \phi(c) = a, \quad \phi(x) = x \quad \text{for all other } x \in A.$$

Both θ, ϕ are bijections and $\phi\theta \neq \theta\phi$ because

$$\phi\theta(a) = \phi(b) = b$$

whereas

$$\theta\phi(a) = \theta(c) = c.$$

The study of groups is central to modern algebra and still very much an active part of research mathematics.

Rings and fields

We have already mentioned rings and fields in chapter 9. They can be described more succinctly using the notion of group and semigroup.

A *ring* consists of a set R and two binary operations $+, .$, such that R is a commutative group under $+$, a semigroup under $.$, and the two operations are related by the distributive laws

$$a(b+c) = ab + ac, \quad (b+c)a = ba + ca \text{ for all } a, b, c \in R.$$

If multiplication is commutative then R is called a *commutative ring*. Thus $\mathbb{Z}$, $\mathbb{Q}$, $\mathbb{R}$, and $\mathbb{C}$ are commutative rings and $\mathbb{H}$ is a (non-commutative) ring. A *field* is a set F with two commutative operations

$+$, . such that F is a group under $+$ with identity 0, $F\backslash\{0\}$ is a group (with identity 1) and the two operations are related by the distributive law

$$a(b+c) = ab + ac \text{ for all } a, b, c \in F.$$

Examples include $\mathbb{Q}$, $\mathbb{R}$, $\mathbb{C}$ but not $\mathbb{Z}$ (because there are non-zero elements without multiplicative inverses) or $\mathbb{H}$ (because multiplication is non-commutative). We have just introduced the notion of a *division ring* D to describe a set with operations $+$, . such that D is a commutative group under $+$ with identity 0, $D\backslash\{0\}$ is a group under . (not necessarily commutative) and the distributive laws hold:

$$a(b+c) = ab + ac, (b+c)a = ba + ca \text{ for all } a, b, c \in D.$$

Examples include $\mathbb{Q}$, $\mathbb{R}$, $\mathbb{C}$, and $\mathbb{H}$.

Vector spaces

Points in three dimensional space can be described by selecting axes and denoting the point by coordinates x, y, z.

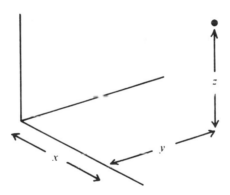

A point in three-dimensional space is just an ordered triple of real numbers (x, y, z). We can regard space as being the set $\mathbb{R}^3$ of ordered triples of real numbers. We can add such triples using the obvious rule:

$$(x_1, y_1, z_1) + (x_2, y_2, z_2) = (x_1+x_2, y_1+y_2, z_1+z_2).$$

Addition is associative and commutative, the triple $(0, 0, 0)$ acts as an identity, and the inverse of (x, y, z) is $(-x, -y, -z)$. Thus $\mathbb{R}^3$ is a commutative group under this addition.

We can also multiply a triple (x, y, z) by an element $a \in \mathbb{R}$ to get

$$a(x, y, z) = (ax, ay, az).$$

This relates to the addition and multiplication on $\mathbb{R}$ according to the rules:

$$(a+b)(x, y, z) = a(x, y, z) + b(x, y, z),$$

$$(ab)(x, y, z) = a(b(x, y, z))$$

$$1(x, y, z) = (x, y, z)$$

and to the addition of vectors as:

$$a\{(x_1, y_1, z_1) + (x_2, y_2, z_2)\} = a(x_1, y_1, z_1) + a(x_2, y_2, z_2).$$

Since we live in three-dimensional space, it seems strange to talk about higher dimensions. Einstein's theory of relativity uses time as a fourth variable, so that a point (x, y, z) at time t is given by the ordered quadruple (x, y, z, t). What is the fifth dimension? The answer is that this approach is a diversion from the mainstream of mathematics. We are certainly constrained to live in three-dimensional space with time as some sort of fourth, but higher dimensions have genuine mathematical significance. For instance to describe the positions of two independent points (x_1, y_1, z_1), (x_2, y_2, z_2), in space requires six real numbers. These can be put in order as a single sextuple $(x_1, y_1, z_1, x_2, y_2, z_2)$ which now describes the position of them both. If we have a rigid body in space, we can describe its position precisely by specifying the positions of three non-collinear points P, Q, R in the body.

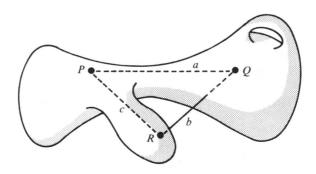

Suppose the distances PQ, QR, RP are a, b, c, then we can place P at the point (x_1, y_1, z_1), move Q to any point (x_2, y_2, z_2) subject only to the restriction that the distance from (x_1, y_1, z_1) to (x_2, y_2, z_2) is a, so†

$$(x_1 - x_2)^2 + (y_1 - y_2)^2 + (z_1 - z_2)^2 = a^2. \qquad (1)$$

Finally we can rotate the body around the axis PQ to put R at a point (x_3, y_3, z_3) subject only to the restrictions $QR = b$, $RP = c$:

$$(x_2 - x_3)^2 + (y_2 - y_3)^2 + (z_2 - z_3)^2 = b^2 \qquad (2)$$

$$(x_3 - x_1)^2 + (y_3 - y_1)^2 + (z_3 - z_1)^2 = c^2. \qquad (3)$$

The position of the rigid body is determined by the nine coordinates x_1, y_1, z_1, x_2, y_2, z_2, x_3, y_3, z_3, subject to the equations (1), (2), (3). It is possible, and by no means bizarre, to consider this as an ordered ninetuple $(x_1, y_1, z_1, x_2, y_2, z_2, x_3, y_3, z_3) \in \mathbb{R}^9$, so that the rigid body's position is given by a point in $\mathbb{R}^9$, subject only to the equations (1), (2), (3). Examples like this in mathematics are legion. Far from being restricted to working in $\mathbb{R}^2$, it is a positive advantage to consider $\mathbb{R}^n$ for $n \in \mathbb{N}$.

Just as in $\mathbb{R}^3$, we can define addition in $\mathbb{R}^n$ and multiplication by real numbers:

$$(x_1, x_2, \ldots, x_n) + (y_1, y_2, \ldots, y_n) = (x_1 + y_1, x_2 + y_2, \ldots, x_n + y_n)$$

$$a(x_1, x_2, \ldots, x_n) = (ax_1, ax_2, \ldots, ax_n).$$

These satisfy the same properties as in $\mathbb{R}^3$. For convenience we will write $v = (x_1, x_2, \ldots, x_n)$, $w = (y_1, y_2, \ldots, y_n)$ then we have:

$$(a + b)v = av + bv$$

$$a(bv) = (ab)v$$

$$1v = v$$

$$a(v + w) = av + aw \quad \text{for all } a, b \in \mathbb{R}, v, w \in \mathbb{R}^n.$$

This is the genesis of the idea of a vector space.

We start with a set V with a binary operation $+$. Then we require a map $m : \mathbb{R} \times V \to V$ where for convenience we will write $m(a, v)$ simply as av. V is said to be a *vector space over* $\mathbb{R}$ if

(VS 1) V is a commutative group under $+$,

---

† Just as the distance between (x_1, y_1), (x_2, y_2) in $\mathbb{R}^2$ is $\sqrt{((x_1 - x_2)^2 + (y_1 - y_2)^2)}$, so the distance between (x_1, y_1, z_1), (x_2, y_2, z_2) in $\mathbb{R}^2$ is $\sqrt{((x_1 - x_2)^2 + (y_1 - y_2)^2 + (z_1 - z_2)^2)}$.

(VS 2) For all $a, b \in \mathbb{R}$, $v, w \in V$,

$$(a+b)v = av + bv$$

$$a(bv) = (ab)v$$

$$1v = v$$

$$a(v+w) = av + aw.$$

This includes the case of $\mathbb{R}^n$, but there are many other interesting cases. For instance let V be the set of all functions from $\mathbb{R}$ to $\mathbb{R}$. Then $f \in V$ means $f: \mathbb{R} \to \mathbb{R}$. We can add together two functions $f, g \in V$ to get $f + g \in V$ simply by defining

$$(f+g)(x) = f(x) + g(x) \text{ for all } x \in \mathbb{R}.$$

For instance if $f(x) = x^3 + x^2$, $g(x) = 3x + 2$, then $(f+g)(x) = x^3 + x^2 + 3x + 2$. Multiplication by $a \in \mathbb{R}$ is given by

$$(af)(x) = a(f(x)) \text{ for all } x \in \mathbb{R}.$$

An example is $f(x) = x^3 + x^2$, $a = -3$, then $(af)(x) = -3(x^3 + x^2)$. The set V is a vector space over $\mathbb{R}$ according to the given definition. In this case the elements of V are functions.

Vector spaces occur in the most unexpected places. Suppose, for example, we try to find the solution $y = f(x)$ of the differential equation

$$\frac{d^2 y}{dx^2} + 95 \frac{dy}{dx} + 1066y = 0.$$

(Here we assume familiarity with calculus.) Then for differentiable function $f: \mathbb{R} \to \mathbb{R}$, $g: \mathbb{R} \to \mathbb{R}$ and real numbers $a, b \in \mathbb{R}$, we find

$$\frac{d}{dx}(af(x) + bg(x)) = a\frac{df(x)}{dx} + b\frac{df(x)}{dx},$$

$$\frac{d^2}{dx^2}(af(x) + bg(x)) = a\frac{d^2 f(x)}{dx^2} + b\frac{d^2 g(x)}{dx^2}.$$

Hence

$$\frac{d^2}{dx^2}(af(x) + bg(x)) + 95\frac{d}{dx}(af(x) + bg(x)) + 1066(af(x) + bg(x))$$

$$= a\left\{\frac{d^2 f(x)}{dx^2} + 95\frac{df(x)}{dx} + 1066f(x)\right\}$$

$$+ b\left\{\frac{d^2 g(x)}{dx^2} + 95\frac{dg(x)}{dx} + 1066g(x)\right\}.$$

If it happens that $y = f(x)$ and $y = g(x)$ are both solutions of the differential equation, then each of the expressions in curly brackets is zero and we can deduce that $y = af(x) + bg(x)$ is also a solution.

Let S be the set of differentiable functions which are solutions of the differential equation. Then (putting $a = b = 1$)

$$f, g \in S \Rightarrow f + g \in S$$

and it is easily seen that S is a commutative group under $+$. Similarly (putting $b = 0$),

$$a \in \mathbb{R}, \ f \in S \Rightarrow af \in S.$$

Checking axioms (VS 1) and (VS 2), we see that the set of solutions S of this differential equation is a vector space over $\mathbb{R}$.

Lest the reader should think our brief description of mathematical structures means that modern algebra is just an arid catalogue of axioms, we should mention some of the striking deductions which have been made from this approach. For two thousand years, since the time of the ancient Greeks, mathematicians tried to trisect an angle using ruler and compass alone. It took an intriguing blend of vector-space theory and field theory to show that the angle 60° (and many others) could not be trisected in this way. (For details, see Clapham [2], or Stewart [16]). In the sixteenth century mathematicians could solve a quadratic equation

$$ax^2 + bx + c = 0$$

by a formula

$$x = \frac{-b \pm \sqrt{(b^2 - 4ac)}}{2a}$$

and had developed more complicated algebraic formulae for any cubic

$$ax^3 + bx^2 + cx + d = 0$$

and any quartic

$$ax^4 + bx^3 + cx^2 + dx + e = 0.$$

For well over two centuries the search continued for an algebraic formula for the solution of a quintic

$$ax^5 + bx^4 + cx^3 + dx^2 + ex + f = 0.$$

In the nineteenth century an intricate chain of deduction using field theory and group theory showed no algebraic formula for the quintic could exist. (See Stewart [16] or Artin [1]).

Various generalizations of the notion of vector space over $\mathbb{R}$ are possible. For instance if we replace $\mathbb{R}$ by a field F, in the definition of a vector space, then we get a *vector space over F*. If we replace it by a ring R, then we get the notion of a *module over R*. The study of these various systems and their applications is the content of modern algebra.

The way ahead

Other constituents of a pure mathematics course in a university today are analysis, geometry, and topology. All these build on the foundations we have described. Analysis is essentially the formal study of limits, differentiation, and integration based on the axioms for the real numbers. At first it concerns itself with functions $f: D \to \mathbb{R}$ where $D \subseteq \mathbb{R}$, then later with functions $f: D \to \mathbb{R}^m$ where $D \subseteq \mathbb{R}^n$. Complex analysis studies differentiable functions $f: D \to \mathbb{C}$ where $D \subseteq \mathbb{C}$. Geometry has been invaded by algebraic structure this century, though retaining a geometrical language to give it the flavour we might expect in the subject. Topology may be considered to have its origins in the concept of a 'metric space'. This is simply a set M with a function $d: M \times M \to \mathbb{R}$ so that $d(x, y)$ behaves like a 'distance function' between x, y. All we require of d is that

  (i) $d(x, y) \geqslant 0$ for all $x, y \in M$.

  (ii) $d(x, y) = 0 \Leftrightarrow x = y$.

  (iii) $d(x, y) = d(y, x)$ for all $x, y \in M$.

  (iv) $d(x, z) \leqslant d(x, y) + d(y, z)$ for all $x, y, z \in M$.

For instance in $\mathbb{R}$, $\mathbb{C}$, or $\mathbb{H}$, we could define $d(x, y) = |x - y|$ and this would satisfy the axioms (i)–(iv), but metric spaces occur in far more general circumstances.

It is not the purpose, nor the intention of this book, to pursue these matters in detail; that is a project which will not be exhausted by an undergraduate degree course, nor yet a lifetime's postgraduate study. Neither is it intended that the student should try to commit any of the algebraic systems to memory: that will come when they are studied more closely later on.

These mathematical ideas may be interpreted in many applications where the structures are appropriate, indeed the applications often dictate the most profitable structures to be studied, applications in physics, engineering, biology, chemistry, economics, statistics, computing, new applications in the social sciences, in psychology, and so on. The purpose of this brief excursion into the broad vista of modern mathematics is to show how the ideas developed in this book form a foundation for the whole of the subject. We now stand on a spring-board, ready to leap into the higher realms of mathematical thought.

Exercises

1. If $z_1, \ldots, z_n$ are complex numbers, prove that
$$|z_1 + \ldots + z_n| \leqslant |z_1| + \ldots + |z_n|.$$

2. Let ω be the complex number defined by $\omega = (-1 + \sqrt{(-3)})/2$. Prove that $\omega^3 = 1$ and that $1 + \omega + \omega^2 = 0$.

3. Prove that $\{1, \omega, \omega^2\}$ is a group under multiplication.

4. Define the concept of an isomorphism $\theta: G \rightarrow H$ where G, H are groups (by analogy with previous definitions). Prove that the group $\{1, \omega, \omega^2\}$ is isomorphic to the group $\mathbb{Z}_3$ under addition.

5. For quaternions p, q verify
   (a) $\overline{p+q} = \bar{p} + \bar{q}$
   (b) $\overline{pq} = \bar{q}\bar{p}$
   (c) $\bar{\bar{q}} = q$
   (d) $q = \bar{q} \Leftrightarrow q \in \mathbb{R}$.

6. For $a, b \in \mathbb{H}$, show $(a+b)^2 = a^2 + ab + ba + b^2$. Give an example to show that we cannot replace this by $(a+b)^2 = a^2 + 2ab + b^2$ in general. If $a \in \mathbb{H}$, $b \in \mathbb{R}$, prove that $(a+b)^2 = a^2 + 2ab + b^2$.
   Solve the equation $x^2 + 2x + 1 = 0$ in $\mathbb{R}$, $\mathbb{C}$, and $\mathbb{H}$. (Let $x = y - 1$ and solve for y.)
   By the substitution $x = y + 1$, solve the equation $x^2 - 2x + 2 = 0$ in $\mathbb{R}$, $\mathbb{C}$, and $\mathbb{H}$.

7. Solve the equation $x(1+j) + k = 2 + i$ for the quaternion x.

8. Solve the equation $ixj + k = 3 + 2j$ for the quaternion x.

9. Find $x, y \in \mathbb{H}$ such that $3ix - 2jy = -1$, $xk + y = 0$.

10. Define *complex quaternions* $\mathbb{H}_\mathbb{C}$ to be quadruples (a_1, a_2, a_3, a_4) of complex numbers, with the same addition and product rules as in the real case. Is $\mathbb{H}_\mathbb{C}$ a division ring?

11. Prove that the complex numbers are *Cauchy complete*, in the following sense:
   If (a_n) is a sequence of complex numbers such that for all $\varepsilon \in \mathbb{R}$, $\varepsilon > 0$, there exists $N \in \mathbb{N}$ such that $|a_m - a_n| < \varepsilon$ for $m, n > N$,
   then (a_n) tends to a limit in $\mathbb{C}$.
   (Hint: Show that $x_n + iy_n \rightarrow x + iy \Leftrightarrow x_n \rightarrow x$ & $y_n \rightarrow y$.)

12. Define a binary operation $\wedge$ on $\mathbb{R}^3$ as follows:
$$(a, b, c) \wedge (d, e, f) = (bf - ce, cd - af, ae - bd).$$

Prove that for all $x, y \in \mathbb{R}^3$,

$$x \wedge y + y \wedge x = 0,$$

$$(x \wedge y) \wedge z + (y \wedge z) \wedge x + (z \wedge x) \wedge y = 0.$$

Is $\wedge$ commutative? Associative?

12

Cardinal numbers

'WHAT IS INFINITY?' When some first-year university students were asked this question recently, the consensus was 'something bigger than any natural number'. In a precise sense, this is correct; one of the triumphs of set theory is that the concept of infinity can be given a clear interpretation. We find not one infinity, but many, a vast hierarchy of infinities. We can answer a question like 'How many rational numbers are there?', with the surprising reply 'as many as there are natural numbers'. The most useful type of question is exemplified by this answer. Rather than ask 'how many' elements there are in a given set, it is much more profitable to compare two sets and ask if there are as many elements in the two of them. This can be described by saying that there are 'the same number of elements' in sets A and B if there is a bijection $f: A \to B$.

Rather than begin with the full hierarchy of infinities, let's begin with what turns out to be the smallest of them. The standard set for comparison purposes we'll take to be the natural numbers $\mathbb{N}$. It is useful to consider $\mathbb{N}$ rather than $\mathbb{N}_0$ simply because a bijection $f: \mathbb{N} \to B$ organizes the elements of B into a sequence; we can call $f(1)$ the *first* element of B using this bijection, $f(2)$ the *second*, and so on . . . Using this process we set up a method of *counting B*. Of course, if we actually say the elements one after another using this bijection, '$f(1), f(2), \ldots$', we never actually reach the end, but we do know that given any element $b \in B$, then $b = f(n)$ for some $n \in \mathbb{N}$, so we will eventually reach that particular element after a certain time.

Recall from chapter 8 that we defined $\mathbb{N}(0) = \varnothing$, and for $n \in \mathbb{N}$,

$$\mathbb{N}(n) = \{m \in \mathbb{N} | 1 \leqslant m \leqslant n\}.$$

We will say that a set X is *finite* if there exists a bijection $f: \mathbb{N}(n) \to X$ for some $n \in \mathbb{N}_0$. We will say that X is *countable* if either X is finite or there exists a bijection $f: \mathbb{N} \to X$. If there's a bijection $f: \mathbb{N}(n) \to X$, then we can say that X has n elements. This is just the usual process of counting. If there's a bijection $f: \mathbb{N} \to X$ then we will say that X has $\aleph_0$ elements. The symbol $\aleph$ is the first letter of the Hebrew alphabet and is called 'aleph', and $\aleph_0$ is the first example we will meet of a new concept of number, used for stating how big an infinite set is. If we say that a set X has $\aleph_0$ elements, it simply means that there is a bijection between $\mathbb{N}$ and X, which is a precise way of saying 'X has the same (cardinal) number of elements as $\mathbb{N}$'. Before we discuss cardinal numbers in general, let's take a closer look at the notion of countability.

EXAMPLE 1. $\mathbb{N}_0$ is countable. Define $f: \mathbb{N} \to \mathbb{N}_0$ by $f(n) = n - 1$, then f is a bijection, This is the first fascinating property of this method of 'counting infinite sets'. $\mathbb{N}$ is a *proper* subset of $\mathbb{N}_0$, so intuitively it should have fewer elements, yet in the sense of a bijection between the sets, they have the same size.

Historically, Galileo gave an even more graphic example in 1638 when he exhibited a correspondence between the natural numbers and the squares of natural numbers:

EXAMPLE 2 (Galileo).   There is a correspondence between the natural numbers and the perfect squares:

$$
\begin{array}{ccccccc}
1 & 2 & 3 & 4 & \ldots & n & \ldots \\
\downarrow & \downarrow & \downarrow & \downarrow & & \downarrow & \\
1 & 4 & 9 & 16 & \ldots & n^2 & \ldots
\end{array}
$$

In modern set theoretic terms, if $S = \{n^2 \in \mathbb{N} \mid n \in \mathbb{N}\}$, the map $f: \mathbb{N} \to S$ given by $f(n) = n^2$ is a bijection.

For over two centuries, this seeming contradiction blighted any attempt to contemplate infinity in a precise sense. Leibniz went as far as to suggest that we should only ever consider finite sets, that the contradiction arose because the natural numbers are infinite. His resolution of the conflict was that if we only considered a finite collection of natural numbers, say the numbers less than a hundred, then there is not a correspondence between the natural numbers under a hundred and the perfect squares which are less than a hundred. Cantor's solution of the paradox in the 1870s was even more dramatic. He showed that if we interpret 'as many' to mean that there is a bijection between two sets, then any infinite set has 'as many' elements as a proper subset! Here 'infinite' is interpreted in the technical sense that B is infinite if there is no bijection $f: \mathbb{N}(n) \to B$ for any $n \in \mathbb{N}_0$.

PROPOSITION 1 (Cantor).   If a set B is infinite, then there exists a proper subset $A \subsetneqq B$ and a bijection $f: B \to A$.

PROOF.   First we select a countably infinite subset X of B. Since no bijection exists between $\mathbb{N}(0)$ and B, B is non-empty and there exists some element in B which we will call x_1. Define a function $g: \mathbb{N} \to B$ inductively by $g(1) = x_1$, and if distinct elements $x_1, x_2, \ldots, x_n$ have been found, then since g cannot give a bijection $g: \mathbb{N}(n) \to B$, there must be another element, which we name $x_{n+1} \in B$, which is distinct from $x_1, \ldots, x_n$; define $g(n+1) = x_{n+1}$. Let

$$X = \{x_n \in B \mid n \in \mathbb{N}\}.$$

Let $A = B \backslash \{x_1\}$, define $f: B \to A$ by

$$f(x_n) = x_{n+1} \quad \text{for } x_n \in X$$

and

$$f(b) = b \qquad \text{for } b \notin X.$$

Then f is a bijection. $\square$

Using Galileo's example, we can be even more dramatic and select a subset C where $B \backslash C$ is *infinite* and yet there is a bijection $f: B \to C$. Just select a countably infinite subset X of B as in the proof, then let

$$C = B \backslash \{x_n \in X \mid n \text{ is not a perfect square}\}$$

and define $f: B \to C$ by

$$f(x_n) = x_{n^2} \quad \text{for } x_n \in X$$

$$f(b) = b \qquad \text{for } b \notin X.$$

Starting with an infinite set B, we can remove a (countably) infinite subset and still be left with a subset C which has 'as many elements' as B!

Cantor's solution to the problem of the infinite was to introduce the concept of a *cardinal number*. For the moment we will assume that for every set X, there is a concept called a cardinal number with the property that if there is a bijection $f: X \to Y$, then X and Y have the same cardinal number, and if there is no bijection, then the cardinal numbers concerned are different. In the case of finite sets, we have a convenient candidate for the cardinal number close at hand. Given a bijection $f: \mathbb{N}(n) \to X$, we can say that the cardinal number of X is n. Likewise, given a bijection $f: \mathbb{N} \to X$, we can say that the cardinal number of X is $\aleph_0$. For other infinite sets we may have to invent new symbols for their cardinal numbers. In general we will denote the cardinal number of X by $|X|$, on the understanding that if there is a bijection $f: X \to Y$, then $|X| = |Y|$ ($|X|$ and $|Y|$ are different symbols for the same cardinal number). If there exists an *injection* $f: X \to Y$, then we say that $|X| \leqslant |Y|$, and, as usual, we may define $|X| < |Y|$ to mean $|X| \leqslant |Y|$ and $|X| \neq |Y|$.

In general, if X is a subset of Y, then the inclusion $i: X \to Y$, $i(x) = x$, is an injection, so we have

$$X \subseteq Y \Rightarrow |X| \leqslant |Y|.$$

Proposition 1 says that for an infinite set B, there exists a proper subset A such that $|A| = |B|$. Thus for infinite sets,

$$X \subsetneqq Y \nRightarrow |X| < |Y|.$$

The dilemma posed by Galileo's example is not so much mathematical as psychological. In extending the system of natural numbers and counting to embrace infinite cardinals, the larger system may not have all the properties of the smaller. Familiarity with the smaller system leads us to expect certain properties and we become confused when all the pieces don't fit as we feel they ought to. Insecurity arose when the square of a complex number violated the real number principle that all squares are positive. This was resolved when we realized that the complex numbers cannot be ordered in the same way as their subset of reals. Likewise we resolve the seeming contradiction of Galileo by seeing that when we interpret 'same cardinal number' in terms of a bijection between sets, then proper inclusion of A in B does not in itself prevent A and B from having the same infinite cardinal.

Now let's return to the notion of countability. In the first place, given any infinite set B, as in the proof of proposition 1, we can select a countably infinite subset $X \subseteq B$. This means that $\aleph_0 = |X| \leqslant |B|$, so $\aleph_0$ is the smallest infinite cardinal. Surprisingly many familiar sets which seem much bigger than $\mathbb{N}$ have cardinality $\aleph_0$.

EXAMPLE 3.   The integers are countable. Define $f: \mathbb{N} \to \mathbb{Z}$ by:

$$f(2n) = n, f(2n-1) = 1-n \text{ for } n \in \mathbb{N},$$

then we get the bijection

$$
\begin{array}{ccccccc}
1 & 2 & 3 & 4 & 5 & 6 & 7 & \ldots \\
\downarrow & \downarrow & \downarrow & \downarrow & \downarrow & \downarrow & \downarrow \\
0 & 1 & -1 & 2 & -2 & 3 & -3 & \ldots \quad .
\end{array}
$$

Notice that although f is a bijection, it doesn't preserve the order (in the sense that $m < n$ does not imply $f(m) < f(n)$, for instance $f(2) > f(3)$). When we set up bijections between sets with an order on them, we may have to do it in a very higgledy-piggledy way, as the next example shows.

EXAMPLE 4.   The rationals are countable. We'll do this in stages, first by counting the positive rationals. A positive rational is p/q, where p and q are natural numbers. One way of counting the rationals is to think of them written out as an array:

$$
\begin{array}{cccccc}
1/1 & 1/2 & 1/3 & 1/4 & \cdots \\
2/1 & 2/2 & 2/3 & 2/4 & \cdots \\
3/1 & 3/2 & 3/3 & 3/4 & \cdots \\
4/1 & 4/2 & 4/3 & 4/4 & \cdots \\
\vdots & \vdots & \vdots & \vdots & \cdots
\end{array}
$$

Now read them off along the 'cross diagonals', first $1/1$, next $1/2, 2/1$, then $1/3, 2/2, 3/1$, and so on:

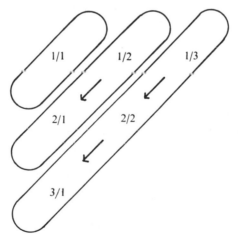

This strings them out as a list $1/1, 1/2, 2/1, 1/3, 2/2, 3/1, \ldots$. In this list there are repeats, because $1/1 = 2/2$ and later on we're going to get $3/3, 4/4$, etc. Similarly $1/2 = 2/4 = 3/6 = \ldots$.Consider each element in the list in turn and delete it if it has occurred before, leaving $1/1$, $1/2, 2/1, 1/3, 3/1, \ldots$. Suppose that the nth rational in the remaining sequence is a_n, then the function f from the natural numbers to the positive rationals $f(n) = a_n$, is a bijection. But every rational is either positive, zero, or negative, so the list $0, a_1, -a_1, a_2, -a_2, \ldots, a_n, -a_n, \ldots$ includes every rational precisely once and the map $g: \mathbb{N} \to \mathbb{Q}$ given by

$$g(1) = 0, \quad g(2n) = a_n, \quad g(2n+1) = -a_n \quad \text{for } n \in \mathbb{N}$$

is a bijection, as required.

It's worth remarking that, although we haven't given an explicit formula for $g(n)$, we have given an explicit prescription for it. The first few terms are

1	2	3	4	5	6	7	8	9	10	11	
↓	↓	↓	↓	↓	↓	↓	↓	↓	↓	↓	
0	1	−1	$\frac{1}{2}$	$-\frac{1}{2}$	2	−2	$\frac{1}{3}$	$-\frac{1}{3}$	3	−3	...

and the reader should be able to continue as far as he wishes. Later on we will develop a far more powerful result (the Schröder–Bernstein Theorem) which will enable us to show two sets have the same cardinal number without having to construct an explicit bijection. By invoking the theorem, this last example can be dealt with much more cleanly.

The reason why we allowed 'countable' to include 'finite' as well as 'countably infinite' is to be found in the next result:

PROPOSITION 2.   A subset of a countable set is countable.

PROOF.   Given a bijection $f: \mathbb{N} \to A$ and $B \subseteq A$, either B is finite, or we can define $g: \mathbb{N} \to B$ by

   $g(1)$ is the least m such that $f(m) \in B$,
   having found $g(1), \ldots g(n)$, then
   $g(n+1)$ is the least m such that $f(m) \in B \backslash \{g(1), \ldots, g(n)\}$.

Informally, this just amounts to writing out the elements of A as a list

$$f(1), f(2), \ldots, f(n), \ldots$$

deleting those terms not in B, leaving the terms in B listed in the same order.   □

The remarkable fact about countable sets is that we can build up sets with them which seem a lot bigger, but once more are countable, in the following precise sense:

PROPOSITION 3.   A countable union of countable sets is countable.

PROOF. Given a countable collection of sets, we can use $\mathbb{N}$ as the index set and write the sets as $\{A_n\}_{n \in \mathbb{N}}$. (If there's only a finite number $A_1, \ldots, A_k$, put $A_n = \varnothing$ for $n > k$.) Since each A_n is countable, we can write the elements of A_n as a list $a_{n1}, a_{n2}, \ldots, a_{nm}, \ldots$ which will terminate if A_n is finite, and be an infinite sequence if A_n is countably infinite. Now tabulate the elements of $\bigcup_{n \in \mathbb{N}} A_n$ as a rectangular array and read them off along the cross diagonals, after the manner of example 4:

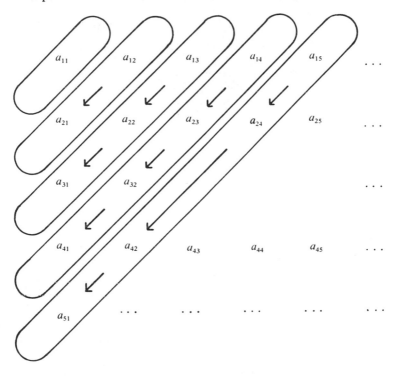

There may be gaps in the array because some of the sets are finite (as in row three of the above illustration) or because there are only a finite number of sets. There may be repeats when two sets A_n, A_m have elements in common, so an element in row n is repeated in row m. We just pass over the gaps and delete elements which have already occurred earlier in the list. The list is then either finite, or an infinite sequence with no repeats. This shows that $\bigcup_{n \in \mathbb{N}} A_n$ is countable. $\square$

PROPOSITION 4.   The cartesian product of two countable sets is countable.

PROOF.   If A and B are countable, write the elements of A as a sequence $a_1, a_2, \ldots, a_n, \ldots$ (which will terminate if A is finite). Similarly write the elements of B as $b_1, b_2, \ldots, b_m, \ldots$. Now write the elements of $A \times B$ as a rectangular array and read them off along the cross diagonals:

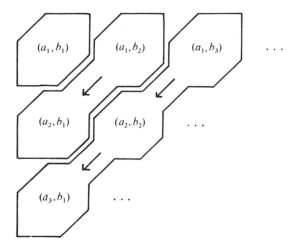

If either A or B is finite, there will be gaps which should be passed over; if both are finite, then $A \times B$ will be finite. If both are infinite, then the explicit bijection $f: \mathbb{N} \to A \times B$ is not hard to write down. There is one element in the first cross diagonal, two in the second, and, in general, n in the nth. So there are $1 + 2 + \ldots + n = \frac{1}{2}n(n+1)$ elements in the first n. The rth element in the next cross diagonal is (a_r, b_{n+2-r}), so the explicit formula for the bijection $f: \mathbb{N} \to A \times B$ is

$$f(m) = (a_r, b_{n+2-r}) \quad \text{for} \quad m = \tfrac{1}{2}n(n+1)+r \quad (1 \leqslant r \leqslant n+1). \quad \square$$

As an instance of proposition 4 in action, we have

EXAMPLE 5.   The set of points in the plane with rational coordinates is countable.

At this stage of the game, the reader may be forgiven for thinking that every infinite set is countable, but that is not so, as we see by looking at the real numbers.

EXAMPLE 6.   The real numbers are not countable. We prove this by contradiction, by showing that no map $f: \mathbb{N} \to \mathbb{R}$ can be surjective, so there cannot be a bijection $f: \mathbb{N} \to \mathbb{R}$. Given a map $f: \mathbb{N} \to \mathbb{R}$, express each $f(m) \in \mathbb{R}$ as a decimal expansion,

$$f(m) = a_m \cdot a_{m1} a_{m2} \ldots a_{mn} \ldots \qquad a_m \in \mathbb{Z}, a_{mn} \in \mathbb{N}_0, 0 \leqslant a_{mn} \leqslant 9$$

where, for definiteness, if the decimal terminates, we'll write it that way, ending in a sequence of zeros, not a sequence of nines. Now we'll write down a real number, different from all the $f(m)$.

Let

$$\beta = 0 \cdot b_1 b_2 \ldots b_n \ldots$$

where

$$b_n = \begin{cases} 1 \text{ if } a_{nn} = 0 \\ 0 \text{ if } a_{nn} \neq 0. \end{cases}$$

Then β is different from $f(n)$ because it differs in the nth place. We've avoided the possible ambiguity which might arise from an infinite sequence of nines in the expansion by making sure that the expansion of β doesn't have any.

Let $\aleph$ be the cardinal number of $\mathbb{R}$. Since $\mathbb{N} \subseteq \mathbb{R}$ we have $\aleph_0 \leqslant \aleph$, and example 6 shows $\aleph_0 \neq \aleph$, so at last we have a cardinal number strictly bigger than $\aleph_0$. In fact for any cardinal number, we can find a strictly bigger one. The cardinal number must be associated with some set A. We simply show that the power set of A has strictly bigger cardinality:

PROPOSITION 5.   If A is a set, then $|\mathbb{P}(A)| > |A|$.

PROOF.   Evidently the map $f: A \to \mathbb{P}(A)$ given by $f(a) = \{a\}$ is an injection, so $|A| \leqslant |\mathbb{P}(A)|$. It remains to show $|A| \neq |\mathbb{P}(A)|$. We simply show that a map $f: A \to \mathbb{P}(A)$ cannot be a surjection. For such a map $f(a) \in \mathbb{P}(A)$ for each $a \in A$, so $f(a)$ is a subset of A. We ask the question 'does a belong to the subset $f(a)$?'. The answer is always 'yes' or 'no' and we select those elements for which the answer is 'no' to get the subset

$$B = \{a \in A \mid a \notin f(a)\}.$$

We now assert that B is not mapped onto by any element of A under the function f, for if B were equal to $f(a)$ for some $a \in A$, the question 'does a belong to B?' leads to a contradiction:

$$a \in B \Rightarrow a \notin f(a)$$

$$a \notin B \Rightarrow a \in f(a)$$

So B is not mapped onto by f and f is not surjective; even more so, it cannot be a bijection. □

Proposition 5 leads us to a hierarchy of infinities. We begin with $\aleph_0 = |\mathbb{N}|$. Then $|\mathbb{P}(\mathbb{N})|$ is strictly bigger, then $|\mathbb{P}(\mathbb{P}(\mathbb{N}))|$, and so on.

The Schröder–Bernstein Theorem

An obvious question to ask concerning the relation $\leqslant$ between cardinals is

if $|A| \leqslant |B|$ and $|B| \leqslant |A|$, can we conclude that $|A| = |B|$?

The answer to this question is in the affirmative and the content of this statement is the Schröder–Bernstein Theorem. The proof is more tricky than at first might seem necessary for such a simple sounding proposition. The main problem is that $|A| \leqslant |B|$ tells us that there is an injection $f : A \to B$ and $|B| \leqslant |A|$ gives us an injection in the opposite direction $g : B \to A$, the only trouble being that these injections needn't be related in any way. Somehow we must use them to construct a bijection between A and B.

THEOREM 6 (Schröder–Bernstein).   Given sets A, B, then $|A| \leqslant |B|$ and $|B| \leqslant |A|$ implies $|A| = |B|$.

PROOF.   We have injections $f : A \to B$, $g : B \to A$. We can use f to pass from A to B or g to pass from B to A. Repeating the process, we

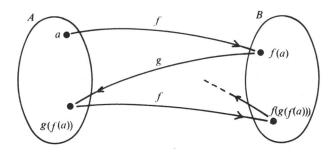

can pass to and fro obtaining $f(a)$, $g(f(a))$, $f(g(f(a)))$, ... The key to the proof is to try to trace such a chain *backwards*. Start with $b \in B$ and see if there exists $a \in A$ such that $f(a) = b$; if such an a exists, it is unique. Then see if there is a $b_1 \in B$ such that $g(b_1) = a$, then $a_1 \in A$ such that $f(a_1) = b_1$, attempting to build up a chain, $b, a, b_1, a_1, \ldots, b_n, a_n$, where $g(b_r) = a_r, f(a_r) = b_{r+1}$. In tracing back a

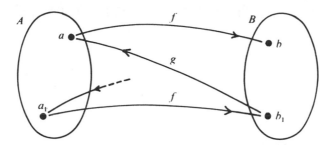

chain of elements in this fashion, three things can happen

(i) we reach $a_N \in A$ and stop because there is no $b^* \in B$ with $g(b^*) = a_N$;

(ii) we reach $b_N \in B$ and stop because no $a^* \in A$ satisfies $f(a^*) = b_N$;

(iii) the process goes on forever.

This partitions B into three sets

(1) B_A, the subset of elements in B whose ancestry originates in A, as in (i)

(2) B_B, the subset of elements in B whose ancestry originates in B, as in (ii).

(3) B_∞, the subset of elements in B whose ancestry can be traced back ad infinitum, as in (iii).

(Note that B_A, B_B, B_∞ are disjoint and their union is B, so they do indeed give a partition.)

Similarly we can partition A into A_A, A_B, A_∞ whose ancestry originates in A, B, or goes back forever, respectively.

It is easily seen that the restriction of f to A_A gives a bijection $f: A_A \to B_A$, the restriction of g to B_B gives a bijection $g: B_B \to A_B$ and the restrictions of f, g both give bijections $f: A_\infty \to B_\infty, g: B_\infty \to A_\infty$. Using the first two and one of the third, we can concoct a bijection $F: A \to B$ by

$$F(a) = \begin{cases} f(a) & \text{if } a \in A_A \\ g^{-1}(a) & \text{if } a \in A_B \\ f(a) & \text{if } a \in A_\infty \end{cases}$$

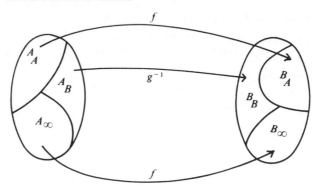

This completes the proof. □

As an example of this theorem in use, we will give an alternative proof of the fact that the rationals are countable. The inclusion $i: \mathbb{N} \to \mathbb{Q}$ shows that $|\mathbb{N}| \leqslant |\mathbb{Q}|$, and, as any rational can be written uniquely in its lowest terms as $(-1)^n p/q$ where $n, p, q \in \mathbb{N}$, by unique factorization the function $f: \mathbb{Q} \to \mathbb{N}$, $f((-1)^n p/q) = 2^n 3^p 5^q$, is an injection, so $|\mathbb{Q}| \leqslant |\mathbb{N}|$.

A more interesting example is to show that $|\mathbb{P}(\mathbb{N})| = \aleph$. An injection $f: \mathbb{P}(\mathbb{N}) \to \mathbb{R}$ can be obtained by

$$f(A) = 0 \cdot a_1 a_2 \ldots a_n \ldots$$

where

$$a_n = \begin{cases} 0 \text{ if } n \notin A \\ 1 \text{ if } n \in A. \end{cases}$$

For each subset $A \subseteq \mathbb{N}$, this gives a unique decimal expansion and f is an injection. To get an injection $g: \mathbb{R} \to \mathbb{P}(\mathbb{N})$ requires a little more cunning. Instead of writing a real number as a decimal expansion, we express it as a *bicimal*, which means that we write it as the limit of fractions of the form

$$a_0 + a_1/2 + a_2/4 + \ldots + a_n/2^n$$

where a_0 is an integer and a_n is 0 or 1 for $n \geqslant 1$. If we exclude such expressions concluding with an infinite sequence of 1s (the bicimal equivalent of the decimal problem involving an infinite sequence of 9s), then such a bicimal expansion is unique. Now express the integer a_0 in binary notion as

$$a_0 = (-1)^m b_k \ldots b_2 b_1$$

where m and the digits $b_1, b_2, \ldots, b_k$ are all 0 or 1, then we have a unique bicimal expansion for each real number x in the form

$$x = (-1)^m b_k \ldots b_2 b_1 \cdot a_1 a_2 \ldots a_n \ldots$$

where m and each digit $b_1, \ldots, b_k, a_1, \ldots, a_n, \ldots$ is 0 or 1. For convenience, in this case write $b_n = 0$ for $n > k$. Now write the terms out as a sequence in the order $m, a_1, b_2, a_2, b_2, \ldots, a_n, b_n, \ldots$. This is a sequence of 0's and 1's and defines a unique subset A of $\mathbb{N}$ according to the rule

'$r \in A$ if and only if the rth term of the sequence is 1'.

In this way we can obtain a function $g \colon \mathbb{R} \to \mathbb{P}(\mathbb{N})$ by defining $g(x)$ to be the subset A determined in this manner. This is an injection, and the Schröder–Bernstein Theorem shows that $|\mathbb{R}| = |\mathbb{P}(\mathbb{N})|$.

Cardinal Arithmetic

Just as we can add, multiply, and take powers of finite cardinal numbers, we can mimic the set-theoretic procedures involved and define corresponding operations on infinite cardinals. Some, but not all, of the properties of ordinary arithmetic carry over to all cardinal numbers, and it is most instructive to see which ones. First of all the definitions:

Addition.   Given two cardinal numbers α, β (finite or infinite), select disjoint† sets A, B such that $|A| = \alpha$, $|B| = \beta$. Define $\alpha + \beta$ to be the cardinal number of $A \cup B$.

Multiplication.   If $\alpha = |A|$, $\beta = |B|$, then $\alpha\beta = |A \times B|$.

Powers.   If $\alpha = |A|$, $\beta = |B|$, then $\alpha^\beta = |A^B|$ where A^B is the set of all functions from B to A.

The reader should pause briefly and check that when the sets concerned are finite then this corresponds to the usual arithmetic. In particular, when $|A| = m$ and $|B| = n$, then on defining a function $f \colon B \to A$, each element $b \in B$ has m possible choices of image, giving m^n functions in all. Addition and multiplication are quite easy in the finite case.

Notice that the sets concerned in the definition of addition have to be disjoint, but this is not necessary in the other two operations. The

† This can always be done. If A and B are not disjoint, replace them by $A' = A \times \{0\}$ and $B' = B \times \{1\}$. Obvious bijections show that $|A'| = |B'|$. Further, A' and B' are disjoint, for if $x \in A' \cap B'$ then $x = (a, 0) = (b, 1)$ so $0 = 1$, a contradiction.

reason for this is obvious: if $|A| = m$, $|B| = n$, and $A \cap B \neq \emptyset$, then $|A \cup B| < m + n$. The most important fact to check about these definitions is that they are indeed well defined. Starting with cardinals α, β, we must choose sets A, B with $|A| = \alpha$, $|B| = \beta$: it is essential to check that if different sets A', B' were used, then the cardinal found in each case would be the same as before. In the case of multiplication, for instance, if $|A| = |A'|$, $|B| = |B'|$, then there are bijections $f: A \to A'$, $g: B \to B'$, which induce a bijection $h: A \times B \to A' \times B'$ given by

$$h(a, b) = (f(a), g(b)).$$

Thus $|A \times B| = |A' \times B'|$, and the product cardinal is well defined. There are corresponding proofs for addition and powers of cardinals.

If we investigate the properties of these arithmetic operations, we find that many properties of finite numbers hold for cardinal numbers in general:

PROPOSITION 7. If α, β, γ are cardinal numbers (finite or infinite), then

    (i) $\alpha + \beta = \beta + \alpha$,
    (ii) $(\alpha + \beta) + \gamma = \alpha + (\beta + \gamma)$,
    (iii) $\alpha + 0 = \alpha$,
    (iv) $\alpha\beta = \beta\alpha$,
    (v) $(\alpha\beta)\gamma = \alpha(\beta\gamma)$,
    (vi) $1\alpha = \alpha$,
    (vii) $\alpha(\beta + \gamma) = \alpha\beta + \alpha\gamma$,
(viii) $\alpha^{\beta + \gamma} = \alpha^\beta \alpha^\gamma$,
    (ix) $\alpha^{\beta\gamma} = (\alpha^\beta)^\gamma$,
    (x) $(\alpha\beta)^\gamma = \alpha^\gamma \beta^\gamma$.

PROOF. Let A, B, C be (disjoint) sets with cardinal numbers α, β, γ, respectively. 0 is the cardinal number of $\emptyset$ and 1 is the cardinal number of any one-element set, say $\{0\}$.

(i)–(iii) follow trivially because $A \cup B = B \cup A$, $(A \cup B) \cup C = A \cup (B \cup C)$ and $A \cup \emptyset = A$. (iv)–(vi) follow because we have obvious bijections, $f: A \times B \to B \times A$ given by $f((a, b)) = (b, a)$, $g: (A \times B) \times C \to A \times (B \times C)$ given by $g(((a, b), c)) = (a, (b, c))$, and $h: \{0\} \times A \to A$ given by $h((0, a)) = a$. (vii) results from the equality $A \times (B \cup C) = (A \times B) \cup (A \times C)$.

If the last three seem harder, it is only an illusion for the reader, caused by the fact that he is less familiar with the set of functions A^B from B to A. We only have to set up the appropriate bijections.

(viii) Define $f: A^{B \cup C} \to A^B \times A^C$ by starting with a map $\phi: B \cup C \to A$, defining $\phi_1: B \to A$ to be the restriction of ϕ to B, $\phi_2: C \to A$ to be the restriction of ϕ to C, then put $f(\phi) = (\phi_1, \phi_2)$. This function f is a bijection.

(ix) Define $g: A^{B \times C} \to (A^B)^C$ by starting with a function $\phi: B \times C \to A$, then defining the function $g(\phi): C \to A^B$ by $[g(\phi)](c): B \to A$ is the function which takes $b \in B$ to

$$([g(\phi)](c))(b) = \phi((b, c)).$$

As this is less familiar, it is worth demonstrating that g is a bijection. It is injective, for if $g(\phi) = g(\psi)$ for two maps ϕ, ψ from $B \times C$ to A, then

$$([g(\phi)](c))(b) = ([g(\psi)](c))(b) \quad \text{for all } b \in B, c \in C$$

so, by definition,

$$\phi((b, c)) = \psi((b, c)) \quad \text{for all } b \in B, c \in C,$$

which means that $\phi = \psi$.

To show that g is surjective, start with a function $\theta \in (A^B)^C$, i.e. $\theta: C \to A^B$, then define $\phi: B \times C \to A$ by

$$\phi(b, c) = [\theta(c)](b) \quad \text{for all } b \in B, c \in C.$$

We have $g(\phi) = \theta$, as required.

(x) The final equality between cardinals follows from the bijection $h: (A \times B)^C \to A^C \times B^C$ given by writing any $\phi: C \to A \times B$ in terms of

$$\phi(c) = (\phi_1(c), \phi_2(c)) \quad \text{for } c \in C,$$

then by setting $h(\phi) = (\phi_1, \phi_2)$. Checking the details is left to the reader. $\square$

Now let's make some explicit calculations with cardinals. As a corollary of proposition 3, we find that

$$n + \aleph_0 = \aleph_0 + n = \aleph_0 \quad \text{for any finite cardinal } n,$$

$$\aleph_0 + \aleph_0 = \aleph_0.$$

This shows us that we have no possibility of defining subtraction of cardinals where infinite cardinals are involved, for what would $\aleph_0 - \aleph_0$ be? According to the above results it could be any finite cardinal or $\aleph_0$ itself, so subtraction cannot be defined uniquely so that

$$\aleph_0 - \aleph_0 = \alpha \Leftrightarrow \aleph_0 = \aleph_0 + \alpha.$$

From proposition 4 it is easy to deduce that

$$n\aleph_0 = \aleph_0 n = \aleph_0 \quad \text{for } n \in \mathbb{N},$$

$$\aleph_0 \aleph_0 = \aleph_0.$$

It is interesting to calculate $0\aleph_0$. This turns out to be zero. In fact we have

$$0\beta = 0 \quad \text{for each cardinal number } \beta.$$

This is because

$$A = \varnothing \Rightarrow A \times B = \varnothing \quad \text{for any other set } B,$$

for if A has no elements, then there are no ordered pairs (a, b) for $a \in A, b \in B$. This means that, in terms of cardinal numbers, zero times infinity is zero, no matter how big the infinite cardinal is.

Likewise it is instructive to calculate α^0 and α^1 for any cardinal α. By definition, if $|A| = \alpha$, then α^0 is the cardinal number of the set of functions from $\varnothing$ to A. You might be forgiven for thinking that there are *no* functions from $\varnothing$ to A, but the set theoretic definition of a function $f: \varnothing \to A$ as a subset of $\varnothing \times A$ exhibits just one function, the empty subset of $\varnothing \times A$. So $\alpha^0 = 1$. Since $|\{0\}| = 1$, α^1 is the cardinal number of the set of functions from $\{0\}$ to A. A function $f: \{0\} \to A$ is uniquely determined by the element $f(0) \in A$, so there is a bijection $g: A^{\{0\}} \to A$ given by $g(f) = f(0)$, showing $|A^{\{0\}}| = |A|$, or $\alpha^1 = \alpha$. By induction using proposition 7 (viii), we get

$$(\aleph_0)^0 = 1, \qquad (\aleph_0)^n = \aleph_0 \quad \text{for } n \in \mathbb{N}.$$

If we calculate 2^α for any cardinal α, we get an interesting result in terms of the power set. Suppose that $|A| = \alpha$, then, since $|\{0, 1\}| = 2$, we have

$$|\{0, 1\}^A| = 2^\alpha.$$

But a function $\phi: A \to \{0, 1\}$ corresponds precisely to a subset of A, namely

$$\{a \in A \,|\, \phi(a) = 1\}.$$

Define $f: \{0, 1\}^A \to \mathbb{P}(A)$ by $f(\phi) = \{a \in A \,|\, \phi(a) = 1\}$, then f is a bijection, so $|\mathbb{P}(A)| = 2^\alpha$. From proposition 5 we see that

$$2^\alpha > \alpha \quad \text{for all cardinal numbers } \alpha.$$

Order of Cardinals

We have already proved a number of results concerning the order of cardinals at various points in this chapter. It is now an opportune moment to collect these together and make the list more comprehensive by filling in the gaps:

PROPOSITION 8.   If $\alpha, \beta, \gamma, \delta$ are cardinal numbers (finite or infinite) then

(i) $\alpha \leq \beta, \beta \leq \gamma \Rightarrow \alpha \leq \gamma$,
(ii) $\alpha \leq \beta, \beta \leq \alpha \Rightarrow \alpha = \beta$,
(iii) $\alpha \leq \beta, \gamma \leq \delta \Rightarrow \alpha + \gamma \leq \beta + \delta$,
(iv) $\alpha \leq \beta, \gamma \leq \delta \Rightarrow \alpha\gamma \leq \beta\delta$,
(v) $\alpha \leq \beta, \gamma \leq \delta \Rightarrow \alpha^\gamma \leq \beta^\delta$.

PROOF.   Select sets A, B, C, D, with cardinals $\alpha, \beta, \gamma, \delta$.

(i) If $f: A \to B$, $g: B \to C$ are injections, then $gf: A \to C$ is an injection.

(ii) Is the Schröder–Bernstein Theorem.

(iii) Given injections $f: A \to B$, $g: C \to D$ where $A \cap B = \varnothing$, $C \cap D = \varnothing$, define $h: A \cup C \to B \cup D$ by

$$h(x) = \begin{cases} f(x) \text{ for } x \in A \\ g(x) \text{ for } x \in C. \end{cases}$$

Since $A \cap B = \varnothing$, this is well defined, and since $B \cap D = \varnothing$, the fact that f, g are injections implies h is an injection.

(iv) Given injections $f: A \to B$, $g: C \to D$, define $p: A \times C \to B \times D$ by

$$p((a, c)) = (f(a), g(c)) \quad \text{for all } a \in A, c \in C.$$

Clearly p is an injection (for if $p((a_1, c_1)) = p((a_2, c_2))$, then $(f(a_1), g(c_1)) = (f(a_2), g(c_2))$, so $f(a_1) = f(a_2)$, $g(c_1) = g(c_2)$, and the injectivity of f, g implies $a_1 = a_2$, $c_1 = c_2$).

(v) This is best visualized by considering $A \subseteq B$, $C \subseteq D$. (If we are given injections $f: A \to B$, $g: C \to D$, simply replace A by $f(A) \subseteq B$, and C by $g(C) \subseteq D$ in the argument which follows.)

For $A \subseteq B$, $C \subseteq D$, to define a map $\mu: A^C \to B^D$, all we need to do is to show how to extend a function $\phi: C \to A$ into a function $\mu(\phi): D \to B$. The easiest way to do this is to select an element $b \in B$, (any one will do, the exceptional case $B = \varnothing$ easily implies (v) by a separate argument);

now define $\mu(\phi) \in B^D$ by

$$[\mu(\phi)](d) = \begin{cases} \phi(d) & \text{for } d \in C, \\ b & \text{for } d \in D \backslash C. \end{cases}$$

Then $\mu: A^C \to B^D$ is an injection because $\mu(\phi_1) = \mu(\phi_2)$ implies

$$[\mu(\phi_1)](d) = [\mu(\phi_2)][(d)] \quad \text{for all } d \in D,$$

in particular, this means that

$$\phi_1(d) = \phi_2(d) \quad \text{for all } d \in C,$$

so $\phi_1 = \phi_2$. $\square$

Looking at this last proposition, the reader will see that there is a notable omission from the list of properties one might expect of an order relation. We have not asserted that any two cardinal numbers are comparable, i.e. that given cardinals α, β, then either $\alpha \leqslant \beta$ or $\beta \leqslant \alpha$. What this would amount to is selecting sets A, B with cardinals α, β respectively and showing that there is either an injection $f: A \to B$, or $g: B \to A$, (or both). To be able to construct such an injection, we would either have to know something about the sets A and B, or we would need some general method of proceeding with the construction of a suitable injection. Given specific sets, we can proceed in an ad hoc fashion and use our ingenuity to try to set up an injection from one to the other. A general method which works for *all* sets requires us to be much more precise about what we mean by a 'set'. It strains the bounds of set theory. Until we put specific restrictions on what we mean by the word 'set' we cannot say how to compare two of them. The theory of sets has grown into a large and living plant; to nourish it we must put down stronger roots into the foundations.

Exercises

1. Let X be the set of points $(x, y, z) \in \mathbb{R}^3$ such that $x, y, z \in \mathbb{Q}$. Is X countable?

2. Let S be the set of spheres in $\mathbb{R}^3$ whose centres have rational coordinates and whose radii are rational. Show that S is countable.

3. Let $[0, 1[$ be the set of real numbers x such that $0 \leqslant x < 1$. By writing each one as a decimal expansion, prove that $[0, 1[$ is uncountable.

4. Which of the following sets are countable? (Prove or disprove each case.)
   (a) $\{n \in \mathbb{N} | n \text{ is prime}\}$,

---

† The function $\mu(\phi): D \to B$ isn't usually an injection; don't confuse this with the function $\mu: A^C \to B^D$.

(b) $\{r \in \mathbb{Q} | r > 0\}$,
(c) $\{x \in \mathbb{R} | 1 < x < 10^{-1,000,000}\}$,
(d) $\mathbb{C}$,
(e) $\{x \in \mathbb{R} | x^2 = 2^a 3^b \text{ for some } a, b \in \mathbb{N}\}$.

5. If $a, b \in \mathbb{R}$ and $a < b$, the *closed interval* $[a, b]$ is

$$[a, b] = \{x \in \mathbb{R} | a \leqslant x \leqslant b\},$$

the *open interval* is $]a, b[= \{x \in \mathbb{R} | a < x < b\}$,
and the *half-open intervals* are

$$[a, b[= \{x \in \mathbb{R} | a \leqslant x < b\},$$
$$]a, b] = \{x \in \mathbb{R} | a < x \leqslant b\}.$$

Prove for $a < b$, $c < d$, that $f : [a, b] \to [c, d]$ given by

$$f(x) = \frac{(b - x)c}{b - a} + \frac{(x - a)d}{b - a}$$

is a bijection. Deduce that any two closed intervals have the same cardinal number.

Prove also that $[a, b]$, $]a, b[$, $[a, b[$, $]a, b]$ all have the same cardinal number. (Hint: Show that $[a, b]$ has the same cardinal number as any one of the other three by choosing c, d such that $a < c < d < b$, and then using the Schröder–Bernstein Theorem.)

6. Prove that the cardinal number of a closed interval, an open interval, and a half-open interval is $\aleph$.

7. Prove that between any two distinct real numbers there are a countable number of rationals and an uncountable number of irrationals.

8. Construct an explicit bijection from $[0, 1]$ to $[0, 1[$. (If all else fails, try using the Schröder–Bernstein construction on the injections $f : [0, 1] \to [0, 1[$, $f(x) = \frac{1}{2}x$, and $g : [0, 1[\to [0, 1]$, $g(x) = x$.)

9. If A_1, A_2 are arbitrary sets, prove

$$|A_1| + |A_2| = |A_1 \cup A_2| + |A_1 \cap A_2|.$$

Generalize to n sets $A_1, \ldots, A_n$.

10. Find counterexamples which demonstrate that the following general statements are *false* for cardinal numbers α, β, γ:
(a) $\alpha < \beta \Rightarrow \alpha + \gamma < \beta + \gamma$,
(b) $\alpha < \beta \Rightarrow \alpha\gamma < \beta\gamma$,
(c) $\alpha < \beta \Rightarrow \alpha^\gamma < \beta^\gamma$,
(d) $\alpha < \beta \Rightarrow \gamma^\alpha < \gamma^\beta$.

11. (a) Define $f : [0, 1[\times [0, 1[\to [0, 1[$ by

$$f((0 \cdot a_1 a_2 \ldots a_n \ldots, 0 \cdot b_1 b_2 \ldots b_n \ldots)) = 0 \cdot a_1 b_1 a_2 b_2 \ldots a_n b_n \ldots.$$

Deduce that $\aleph^2 = \aleph$.

(b) Prove the result of (a) more elegantly by using $2^{\aleph_0} = \aleph$, and the properties of cardinal arithmetic.

(c) Using $1.\aleph \leqslant \aleph_0 \aleph \leqslant \aleph\aleph$, or otherwise, find $\aleph_0 \aleph$.

(d) What is $n\aleph$ for $n \in \mathbb{N}$?

(e) Prove $\aleph^{\aleph_0} = \aleph$ and $\aleph^\aleph = 2^\aleph$.

(f) Find $\aleph_0^\aleph$.

12. Given an infinite cardinal α, it may be shown that there exists a cardinal number β such that $\alpha = \aleph_0 \beta$. Use this fact to show that $\aleph_0 \alpha = \alpha$.

13. (The proof by which Cantor showed that there exist transcendental numbers without actually specifying any!)

A real number is said to be *algebraic* if it is the solution if a polynomial equation

$$a_n x^n + \ldots + a_1 x + a_0 = 0$$

with integer coefficients. If not it is said to be *transcendental*.

(a) Show that the set of polynomials with integer coefficients is countable.

(b) Show that the set of algebraic numbers is countable.

(c) Show that some real numbers must be transcendental.

(d) How many transcendental numbers are there?

PART IV

Strengthening the foundations

CHAPTERS 11 AND 12 have shown how the material developed so far can lead into the main body of mathematics, into ever higher realms. But the final chapter of this book will lead in the opposite direction, down into the depths.

For, having constructed such a fine building, it becomes prudent to re-examine the ground on which it rests. What we have done is to replace a very complicated system of intuitions about numbers by a rather simpler system of intuitions about *sets*. Our set-theoretic basis is still intuitive, and informal. If we had built a bungalow, this might not have been important; but we have built a skyscraper (and one that can be extended to much greater heights), and it is time we dug a little deeper into the foundations to see whether they can support the weight. Or, in horticultural terms, we must make sure that the roots of our plant will support the fully grown organism, improve the soil, use higher quality fertilizer, and better seed.

Our aim will be to indicate what *can* be done, and not actually to do it; so we shall talk in an informal way about the possibility of finding a system of axioms for set theory itself.

It may seem that the argument has come full circle: here we are right back at the beginning, worrying about the same kind of things as before. In fact this is not so: we have come more in a *spiral*, returning to the same point but at a higher level. We understand the problems involved, and their solutions, much better than before. The material we have covered so far is quite adequate for almost all of a university course in mathematics. But we should not be so stupid as to imagine that we have reached a complete and final solution, or that total perfection has now been attained.

13

Axioms for set theory

Up to this point, we have concentrated on deriving a formal structure for arithmetic, on the basis of set theory. This analysis should have given us a deepened understanding of the various number systems, the way they work, and their place in the scheme of things. It should also have sharpened the reader's critical faculties and his appreciation for logical rigour. It may have sharpened them sufficiently for him to see that one fundamental ingredient is still lacking. We have axiomatized everything that we can lay hands on, with one notable exception: set theory itself.

Having taken such pains with the structural detail of the number systems, it would be a great pity if the basis on which we have worked should turn out to be defective, and unable to support the weight of the superstructure erected on it. In the ultimate analysis, it is hardly more satisfactory to base a formal theory of numbers on an informal, intuitive, and naive theory of sets; than it is to start with an informal, intuitive, and naive theory of numbers themselves!

However, we may yet escape this otherwise devastating criticism by returning to our starting point and axiomatizing set theory as well. (It would, indeed, have been pleasant to have started off from an axiomatic basis for set theory, except that there are enormous psychological barriers involved in doing something so far removed from reality with no idea why it is needed!) We shall not go into the details very deeply (the reader who wants to see them should consult Mendelson [10]), nor shall we adopt an overly formal style in discussing them. Our aim is just to make clear the unconscious assumptions that have been made about sets, to discard some over-optimistic ones which lead to paradoxes, and to list a system of axioms which appears to be satisfactory.

Historically, some mathematicians hoped for more than this. At the turn of the century a number of them, led by David Hilbert, embarked upon a kind of Arthurian Quest for Truth: a firm and immutable basis for mathematics, and a guarantee that the truths of mathematics could be rendered absolute. In this impermanent and uncertain universe, it is hardly surprising that the Holy Grail turned out, in the end, to be a Mare's Nest.

Some difficulties

The problems with naive set theory are of two kinds. First there are the *paradoxes*: apparently contradictory results obtained by apparently impeccable logic. Then there are purely technical difficulties: are infinite cardinals always comparable? Is there a cardinal between $\aleph_0$ and $2^{\aleph_0}$?

By way of motivation, we consider two paradoxes. The first, due to Bertrand Russell, has been alluded to in chapter 3. If

$$S = \{x \mid x \notin x\},$$

then is $S \in S$ or $S \notin S$? Either answer directly implies the other!

For the second, let U be the set of *all* things, defined (say) by

$$U = \{x \mid x = x\}.$$

Now $X \subseteq U$ for every set U. In particular the power set $\mathbb{P}(U) \subseteq U$. But now, taking cardinals, we have

$$|U| \geq |\mathbb{P}(U)|,$$

but by proposition 5 of chapter 12,

$$|U| < |\mathbb{P}(U)|.$$

This is a contradiction: what's wrong?

Many responses are possible, for example:

The Ostrich.   Ignore the difficulties and maybe they'll go away.

The Drop-out.   The paradoxes point to unavoidable defects in mathematics. Give up, and take up something more profitable such as knitting or sociology.

The Optimist.   Re-examine the reasoning, isolate the source of the difficulties, and try to salvage what is worth saving while disposing of the paradoxes.

If you agree with the Ostrich: stop reading here. If with the Drop-out; burn this book. If with the Optimist: read on ...

Sets and classes

We shall, in the next few sections, discuss one possible solution to the problems, known as *von Neumann–Bernays–Gödel set theory*. This starts from the observation that a plausible source of trouble is the freedom to form weird, and very large, sets (for example, the two sets S and U defined above). All of the known paradoxes seem to 'cheat' in this kind of way.

We therefore distinguish two things. *Classes*, which may be thought of as arbitrary collections (what we hitherto have naively called 'sets'), and *sets*, which are respectable sorts of classes. Then we restrict our ability to define weird or large creatures to classes only. This is the idea: the details are roughly as follows.

Classes are introduced as a primitive, undefined term, along with a relation $\in$ (corresponding to our intuitive idea of membership) and its negation $\notin$. If X and Y are classes then one or other of

$$X \in Y, \qquad X \notin Y$$

is required to hold. We define equality of classes $X = Y$ by

$$(\forall Z)(Z \in X \Leftrightarrow Z \in Y).$$

We say that a class X is a *set* if $X \in Y$ for some class Y. This is the crucial definition: sets are those things which can be *members* of other things. This is in contradistinction to the intuitive feeling that sets are things of which other things are members; and the difference is what makes it hard to define weird and large sets. To make this work, we agree that an expression like

$$\{x \mid P(x)\}$$

shall mean 'the class of all *sets* x for which $P(x)$ is true'. This restriction is forced upon us, because only sets can be members of classes anyway; it has the effect of blocking paradoxes.

For example consider Russell's class

$$S = \{X \mid X \notin X\}.$$

In our new interpretation, this is the class of all *sets* X such that $X \notin X$. Let us run through the usual argument for a contradiction, and see what happens. Suppose $S \in S$. Then S is a member of something, so is a set, so $S \notin S$, a contradiction. Now suppose $S \notin S$. If S is a set, then it satisfies the defining property $X \notin X$, so by the definition, $S \in S$. This is a contradiction too. There remains, however, the possibility that S is *not* a set. In this case we *cannot* deduce that $S \in S$: elements of S have to be sets as well as not being members of themselves.

The upshot of it all is that we don't get a paradox: all we get is a proof that S is not a set. Classes which are not sets are called *proper classes*;

we have just proved they exist. Similarly U may be proved a proper class, and again there is no paradox.

The axioms themselves

The majority of the axioms required come as an anticlimax, because all they do is express that things we obviously want to be sets *are* sets. For convenience in stating them, we assume that the usual notation of set theory has been set up, in the obvious way, for classes. for instance we define

$$\varnothing = \{x \mid x \neq x\}$$

$$\{x, y\} = \{u \mid x = u \text{ or } y = u\}$$

and so on.

From now on we make the convention that small letters $x, y, z, \ldots$ stand for sets, whereas capitals $X, Y, Z, \ldots$ stand for classes, which may or may not be sets.

(S1) *Extensionality* $X = Y \Leftrightarrow (\forall Z)(X \in Z \Leftrightarrow Y \in Z)$.

(We have defined equality of classes as 'having the same members'. This purely technical axiom says that equal classes belong to the same things.)

(S2) *Null set.* $\varnothing$ is a set.

(S3) *Pairs.* $\{x, y\}$ is a set for all sets x, y.

We can now define *singletons* by $\{x\} = \{x, x\}$, then ordered pairs using the Kuratowski definition $(x, y) = \{\{x\}, \{x, y\}\}$, then functions, relations, as before.

(S4) *Membership.* $\in$ is a relation, that is, there exists a class M of ordered pairs (x, y) such that $(x, y) \in M \Leftrightarrow x \in y$.

(S5) *Intersection.*   If X, Y are classes, there is a class $X \cap Y$.

(S6) *Complement.*   If X is a class, its complement X^c exists and is a class.

(S7) *Domain.*   If X is a class of ordered pairs, there exists a class Z such that $u \in Z \Leftrightarrow (u, v) \in X$ for some v.

Much more interesting is an axiom for the definition of a class by a property of its elements, analogous to our $\{x \mid P(x)\}$. We state here a general axiom: it can be derived if desired from a small number of more specialized axioms of the same type.

(S8) *Class existence.*   Let $\phi(X_1, \ldots, X_n, Y_1, \ldots, Y_m)$ be a compound predicate statement in which only set variables are quantified.

Then there exists a class Z such that

$$(x_1, \ldots, x_n) \in Z \Leftrightarrow \phi(x_1, \ldots, x_n, Y_1, \ldots, Y_m).$$

We write

$$Z = \{(x_1, \ldots, x_n) | \phi(x_1, \ldots, x_n, Y_1, \ldots, Y_m)\}.$$

Notice that the xs here are *sets*. In particular, the class

$$Z = \{x | P(x)\}$$

contains as members only those *sets* x for which $P(x)$ is true. This, as we saw above, allows us to avoid paradoxes.

(S9) *Union.*   The union of a set of sets is a set.

(S10) *Power set.*   If x is a set, so is $\mathbb{P}(x)$.

(S11) *Subset.*   If x is a set and X a class, then $x \cap X$ is a set.

Finally, there is an axiom which asserts a slight generalization of the following:

(S12) *Replacement.*   If f is a function whose domain is a set, then its image is a set.

These axioms suffice for almost all of the constructions we have made using set theory. However, they all hold good even if we restrict ourselves only to *finite* sets. We therefore need an axiom to say that infinite sets exist, otherwise we cannot construct any of our beloved number systems. We therefore add an axiom introduced in chapter 8 (von Neumann's brainwave):

(S13) *Axiom of infinity.*   There exists a set x such that $\emptyset \in x$, and whenever $y \in x$ it follows that $y \cup \{y\} \in x$.

Using von Neumann's definition of natural numbers, this axiom boils down to the assertion that the natural numbers form a *set*. It is pretty clear that without some such assertion, set theory would not be much use.

The thirteen axioms listed so far suffice for almost all of our previous work, though a detailed proof is (as usual) somewhat involved and tedious. However, some of the problems in the chapter on cardinals require more delicate axioms yet.

The axiom of choice

In chapter 12, proposition 1, we used an argument which involved selecting an element x_1 from a set B, then x_2 from $B \backslash \{x_1\}, \ldots$, and in general an element x_{n+1} from $B \backslash \{x_1, \ldots, x_n\}$. Although this looks like a recursion argument, it is not covered by theorem 2 of chapter 8, since x_{n+1} is found by an arbitrary choice and not in terms of a previously specified function. Roughly speaking, the method asks us to make 'infinitely many arbitrary choices'. It turns out (though not easily!) that the list of axioms we have so far produced is insufficient to justify this. We therefore state an additional axiom:

(S14) *Axiom of choice.* If $\{x_\alpha\}_{\alpha \in a}$ is an indexed family of sets (with an index *set a*) then there exists a function f such that

$$f: a \to \bigcup_{\alpha \in a} x_\alpha$$

and

$$f(\alpha) \in x_\alpha \quad \text{for each } \alpha \in a.$$

In other words, f 'chooses' for each $\alpha \in a$ an element of x_α.

The status of this axiom is intuitively hard to grasp, but is now well understood. Neither its truth nor its falsity contradict axioms (S1)–(S13) (in the same way that neither the truth nor the falsity of the commutative law contradict the axioms for a group: there exist both commutative and non-commutative groups). The first fact was proved by Kurt Gödel in 1940, the second (a long unsolved problem) by Paul Cohen in 1963. For this reason it is customary in mathematics to point out whenever the axiom of choice is being used, whereas the ordinary axioms (S1)–(S13) are not normally mentioned.

Assuming the axiom of choice allows us to tidy up one loose end. It implies that for any sets x, y, we have either $|x| \geqslant |y|$ or $|y| \geqslant |x|$. (For a proof see Mendelson [10] p. 198.)

The *Continuum hypothesis* that there is no infinite cardinal lying properly between $\aleph_0$ and $2^{\aleph_0}$ goes the same way: neither its truth nor its falsity contradicts (S1)–(S13), or even (S1)–(S14). The proofs are again due to Gödel and Cohen. It is perhaps surprising that so specific a problem should have such an unspecific answer; but it shows how delicate the problems are.

Other, different axioms have also been proposed at various times; and many of the relations between them are now understood quite well. We refer the reader to more specialized texts.

Consistency

However, there is one final problem we must face. Having got our set of axioms, how do we *know* that no paradoxes arise? We certainly seem to have avoided them (for instance, no-one has ever been able to find any), but how can we be *certain* there are no hidden contradictions? A firm, final answer to this question is now known. Unfortunately, this is it: we can *never* be certain.

To explain this, we must go back to the time of David Hilbert. Let us call a system of axioms *consistent* if it does not lead to contradictions. Hilbert wanted to prove that the axioms for set theory were consistent.

For some axiom systems this is easy. If we can find a *model* for the axioms, that is, a structure which satisfies them, then they must be consistent (or else the model could not exist). The trouble is, what materials do we allow for the construction of the model? It is generally agreed that a *finite* model is unexceptionable, because any assertion about it can be checked, in principle, in a finite time. But the axiom of infinity, for example, means that we cannot find a finite model for set theory.

Hilbert's idea was that something less restricted would suffice: what he called a *decision procedure*. This is, so to speak, a finite computer programme which, when fed a formula in set theory, runs through a process to decide whether or not it is true (like the truth-table method for propositions). If we can find such a programme, and *prove* that it always works, then we can feed it the equation

$$0 \neq 0$$

and see what it says. If it says 'true' then our axioms must be *in*consistent, since it can be shown that any contradiction implies the above proposition (use a vacuous argument by contradiction: *anything* is true in an inconsistent system!). And for a while it looked as if Hilbert's idea might work.

Then Kurt Gödel dashed all hopes, by proving two theorems. The first is that there exist in set theory theorems which are true, but for which there neither exists a proof nor a disproof. The second: that if set theory is consistent, then there does not exist any decision procedure which will prove it so.

The proofs of Gödel's theorems are quite technical: they are sketched in Stewart [15] p. 294–5. But they demolish Hilbert's hope of a complete consistency proof.

Does this mean that it is, after all, pointless to seek greater logical rigour in mathematics? After all, if at the end the whole thing hovers in limbo, it hardly seems worth bothering in the first place. This is emphatically *not* the moral to be drawn. Without a proper search for rigour, we would never have reached Gödel's theorems. What they do is pin down certain problems inherent in the axiomatic approach itself.

They do not demonstrate it to be futile: on the contrary, it provides an adequate framework for the whole of modern mathematics, and an inspiration for the development of new ideas. But with Gödel's theorems we can avoid deluding ourselves that everything is perfect, and understand the limitations of the axiomatic method as well as its strengths.

Exercises

1. Show that the axiom of choice implies that if $f: A \to B$ is a surjection, then $|A| \geq |B|$. Conversely, in the context of the other axioms of set theory, prove that the latter fact implies the axiom of choice.

2. Given a collection of sets $\{X_\alpha\}_{\alpha \in A}$ indexed by a set A, the *cartesian product* is defined to be the set of all functions $f: A \to \bigcup_{\alpha \in A} X_\alpha$ such that $f(\alpha) \in X_\alpha$. Show that for $A = \{1, 2, \ldots, n\}$ this corresponds to the usual definition of $X_1 \times X_2 \times \ldots \times X_n$.

Prove that the axiom of choice is equivalent to the assertion that every cartesian product of non-empty sets is itself non-empty.

3. Show that there is a choice involved in the proof of proposition 1 of chapter 12. Express it in terms of a function from a set of subsets of B to B. Is it necessary in this case to include all the subsets of B in the choice?

4. Reconsider Goldbach's conjecture (exercise 13 at the end of chapter 8), which postulates that every positive even integer is the sum of two primes. Look at as many cases of this as you wish to see if there is any pattern to the primes which occur. Convince yourself that Goldbach's conjecture might be true but there may be no single proof which will work for every case. On the other hand, there is always the possibility that the conjecture is false for some very large integer, which we have not yet found.

5. Given a predicate $P(n)$ valid for all $n \in \mathbb{N}$, such that a proof for each $P(n)$ exists in a finite number of lines as explained in chapter 6, is it reasonable to expect that there is a proof of

$$\forall n \in \mathbb{N}: P(n)$$

in this sense?

6. Read the preface to this book again and the introduction to each of the four parts into which the book is divided. Now review the exercises at the end of chapter one. If you still have the solutions that you wrote out at the time you first read chapter one, so much the better. If the book has achieved its purpose, your view on many of these topics will have matured and changed. You should now be in a position to appreciate the kind of thinking used in more advanced mathematics, together with an idea of the sort of problems in the foundations of the subject which are worthy of further study.

References

WE list below a selection of books suitable for further reading (and including those referred to in the text). Those which require extra mathematical background are marked with an asterisk.

1.* ARTIN, E. (1959). *Galois theory*. Notre Dame, Indiana.
2. CLAPHAM, C. R. J. (1969). *Introduction to abstract algebra*. Routledge & Kegan Paul.
3. COXETER. H. S. M. (1961) *Introduction to geometry*. Wiley.
4. FOWLER, D. H. (1973). *Introducing real analysis*. Transworld Student Library.
5. GREEN, J. A. (1965). *Sets and groups*. Routledge & Kegan Paul.
6. —— (1958). *Sequences and series*. Routledge & Kegan Paul.
7. HALMOS, P. R. (1960). *Naive set theory*. Van Nostrand.
8. HILTON, P. J. and GRIFFITHS, H. B. (1970). *A comprehensive textbook of classical mathematics: a contemporary interpretation*. Van Nostrand Reinhold.
9. KLINE, M. (1969). *Mathematics in the modern world: readings from Scientific American*. Freeman.
10.* MENDELSON, E. (1964). *Introduction to mathematical logic*. Van Nostrand.
11. MUNKRES, J. R. (1964). *Elementary linear algebra*. Addison-Wesley.
12. NAGEL, E. and NEWMAN, J. R. (1958). *Gödel's proof*. Routledge & Kegan Paul.
13. SKEMP, R. R. (1971). *The psychology of learning mathematics*. Penguin.
14. SPIVAK, M. (1967). *Calculus*. Benjamin.
15. STEWART, I. N. (1975). *Concepts of modern mathematics*. Penguin.
16.* —— (1973). *Galois theory*. Chapman & Hall.

Index